EXCLUSIVE
EMPLOYMENT ADVICE

BRICKLAYING

LEVEL 1 DIPLOMA

JOHN CARRUTHERS

IAN COOTE

Nelson Thornes

Published in 2013 by:

Nelson Thornes Ltd
Delta Place
27 Bath Road
CHELTENHAM
GL53 7TH

United Kingdom

13 14 15 16 17 / 10 9 8 7 6 5 4 3 2 1

A catalogue record for this book is available from the British Library

ISBN 978 1 4085 2122 9

Cover photograph: DNY59/iStockphoto

Page make-up by GreenGate Publishing Services, Tonbridge, Kent

Printed in Croatia by Zrinski

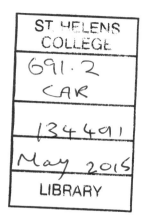
Note to learners and tutors

This book clearly states that a risk assessment should be undertaken and the correct PPE worn for the particular activities before any practical activity is carried out. Risk assessments were carried out before photographs for this book were taken and the models are wearing the PPE deemed appropriate for the activity and situation. This was correct at the time of going to print. Colleges may prefer that their learners wear additional items of PPE not featured in the photographs in this book and should instruct learners to do so in the standard risk assessments they hold for activities undertaken by their learners. Learners should follow the standard risk assessments provided by their college for each activity they undertake which will determine the PPE they wear.

CONTENTS

INTRODUCTION

About this book

This book has been written for the Cskills Awards Level 2 Diploma in Bricklaying. It covers all the units of the qualification, so you can feel confident that your book fully covers the requirements of your course.

This book contains a number of features to help you acquire the knowledge you need. It also demonstrates the practical skills you will need to master to successfully complete your qualification. We've included additional features to show how the skills and knowledge can be applied to the workplace, as well as tips and advice on how you can improve your chances of gaining employment.

The features include:

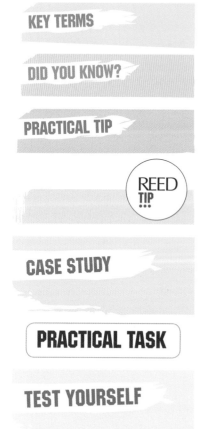

* chapter openers which list the learning outcomes you must achieve in each unit

* key terms that provide explanations of important terminology that you will need to know and understand

* Did you know? margin notes to provide key facts that are helpful to your learning

* practical tips to explain facts or skills to remember when undertaking practical tasks

* Reed tips to offer advice about work, building your CV and how to apply the skills and knowledge you have learnt in the workplace

* case studies that are based on real tradespeople who have undertaken apprenticeships and explain why the skills and knowledge you learn with your training provider are useful in the workplace

* practical tasks that provide step-by-step directions and illustrations for a range of projects you may do during your course

* Test yourself multiple choice questions that appear at the end of each unit to give you the chance to revise what you have learnt and to practise your assessment (your tutor will give you the answers to these questions).

Further support for this book can be found at our website, www.planetvocational.com/subjects/build

CONTRIBUTORS TO THIS BOOK

Reed Property & Construction

Reed Property & Construction specialises in placing staff at all levels, in both temporary and permanent positions, across the complete lifecycle of the construction process. Our consultants work with most major construction companies in the UK and our clients are involved with the design, build and maintenance of infrastructure projects throughout the UK.

Expert help

As a leading recruitment consultancy for mid–senior level construction staff in the UK, Reed Property & Construction is ideally placed to advise new workers entering the sector, from building a CV to providing expertise and sharing our extensive sector knowledge with you. That's why, throughout this book, you will find helpful hints from our highly experienced consultants, all designed to help you find that first step on the construction career ladder. These tips range from advice on CV writing to interview tips and techniques, and are all linked in with the learning material in this book.

Work-related advice

Reed Property & Construction has gained insights from some of our biggest clients – leading recruiters within the industry – to help you understand the mind-set of potential employers. This includes the traits and skills that they would like to see in their new employees, why you need the skills taught in this book and how they are used on a day to day basis within their organisations.

Getting your first job

This invaluable information is not available anywhere else and is all geared towards helping you gain a position with an employer once you've completed your studies. Entry level positions are not usually offered by recruitment companies, but the advice we've provided will help you to apply for jobs in construction and hopefully gain your first position as a skilled worker.

CONTRIBUTORS TO THIS BOOK

The case studies in this book feature staff from Laing O'Rourke and South Tyneside Homes.

Laing O'Rourke is an international engineering company that constructs large-scale building projects all over the world. Originally formed from two companies, John Laing (founded in 1848) and R O'Rourke and Son (founded in 1978) joined forces in 2001.

At Laing O'Rourke, there is a strong and unique apprenticeship programme. It runs a four-year 'Apprenticeship Plus' scheme in the UK, combining formal college education with on-the-job training. Apprentices receive support and advice from mentors and experienced tradespeople, and are given the option of three different career pathways upon completion: remaining on site, continuing into a further education programme, or progressing into supervision and management.

The company prides itself on its people development, supporting educational initiatives and investing in its employees. Laing O'Rourke believes in collaboration and teamwork as a path to achieving greater success, and strives to maintain exceptionally high standards in workplace health and safety.

South Tyneside Homes was launched in 2006, and was previously part of South Tyneside Council. It now works in partnership with the council to repair and maintain 18,000 properties within the borough, including delivering parts of the Decent Homes Programme.

South Tyneside Council's Housing Company

South Tyneside Homes believes in putting back into the community, with 90 per cent of its employees living in the borough itself. Equality and diversity, as well as health and wellbeing of staff, is a top priority, and it has achieved the Gold Status Investors in People Award.

South Tyneside Homes is committed to the development of its employees, providing opportunities for further education and training and great career paths within the company – 80 per cent of its management team started as apprentices with the company. As well as looking after its staff and their community, the company looks after the environment too, running a renewable energy scheme for council tenants in order to reduce carbon emissions and save tenants money.

The apprenticeship programme at South Tyneside Homes has been recognised nationally, having trained over 80 young people in five main trade areas over the past six years. One of the UK's Top 100 Apprenticeship Employers, it is an Ambassador on the panel of the National Apprentice Service. It has won the Large Employer of the Year Award at the National Apprenticeship Awards and several of its apprentices have been nominated for awards, including winning the Female Apprentice of the Year for the local authority.

Chapter 1

Unit CSA–L1Core01

HEALTH, SAFETY AND WELFARE IN CONSTRUCTION AND ASSOCIATED INDUSTRIES

LEARNING OUTCOMES

LO1: Know the health and safety regulations, roles and responsibilities

LO2: Know the accident and emergency procedures and how to report them

LO3: Know how to identify hazards on construction sites

LO4: Know about health and hygiene in a construction environment

LO5: Know how to handle and store materials and equipment safely

LO6: Know about basic working platforms and access equipment

LO7: Know how to work safely around electricity in a construction environment

LO8: Know how to use personal protective equipment (PPE) correctly

LO9: Know the fire and emergency procedures

LO10: Know about signs and safety notices

INTRODUCTION

The aim of this chapter is to:

* help you to source relevant safety information

* help you to use the relevant safety procedures at work.

HEALTH AND SAFETY REGULATIONS, ROLES AND RESPONSIBILITIES

The construction industry can be dangerous, so keeping safe and healthy at work is very important. If you are not careful, you could injure yourself in an accident or perhaps use equipment or materials that could damage your health. Keeping safe and healthy will help ensure that you have a long and injury-free career.

Although the construction industry is much safer today than in the past, more than 2,000 people are injured and around 50 are killed on site every year. Many others suffer from long-term ill-health such as deafness, spinal damage, skin conditions or breathing problems.

Key health and safety legislation

Laws have been created in the UK to try to ensure safety at work. Ignoring the rules can mean injury or damage to health. It can also mean losing your job or being taken to court.

The two main laws are the Health and Safety at Work etc. Act (HASAWA) and the Control of Substances Hazardous to Health Regulations (COSHH).

The Health and Safety at Work etc. Act (HASAWA) (1974)

This law applies to all working environments and to all types of worker, sub-contractor, employer and all visitors to the workplace. It places a duty on everyone to follow rules in order to ensure health, safety and welfare. Businesses must manage health and safety risks, for example by providing appropriate training and facilities. The Act also covers first aid, accidents and ill health.

Reporting of Injuries, Diseases and Dangerous Occurrences Regulations (RIDDOR) (1995)

Under RIDDOR, employers are required to report any injuries, diseases or dangerous occurrences to the Health and Safety Executive (HSE). The regulations also state the need to maintain an accident book.

Control of Substances Hazardous to Health (COSHH) (2002)

In construction, it is common to be exposed to substances that could cause ill health. For example, you may use oil-based paints or preservatives, or work in conditions where there is dust or bacteria.

Employers need to protect their employees from the risks associated with using hazardous substances. This means assessing the risks and deciding on the necessary precautions to take.

Any control measures (things that are being done to reduce the risk of people being hurt or becoming ill) have to be introduced into the workplace and maintained; this includes monitoring an employee's exposure to harmful substances. The employer will need to carry out health checks and ensure that employees are made aware of the dangers and are supervised.

Control of Asbestos at Work Regulations (2012)

Asbestos was a popular building material in the past because it was a good insulator, had good fire protection properties and also protected metals against corrosion. Any building that was constructed before 2000 is likely to have some asbestos. It can be found in pipe insulation, boilers and ceiling tiles. There is also asbestos cement roof sheeting and there is a small amount of asbestos in decorative coatings such as Artex.

Asbestos has been linked with lung cancer, other damage to the lungs and breathing problems. The regulations require you and your employer to take care when dealing with asbestos:

* You should always assume that materials contain asbestos unless it is obvious that they do not.

* A record of the location and condition of asbestos should be kept.

* A risk assessment should be carried out if there is a chance that anyone will be exposed to asbestos.

The general advice is as follows:

* Do not remove the asbestos. It is not a hazard unless it is removed or damaged.

* Remember that not all asbestos presents the same risk. Asbestos cement is less dangerous than pipe insulation.

* Call in a specialist if you are uncertain.

Provision and Use of Work Equipment Regulations (PUWER) (1998)

PUWER concerns health and safety risks related to equipment used at work. It states that any risks arising from the use of equipment must either be prevented or controlled, and all suitable safety measures must have been taken. In addition, tools need to be:

* suitable for their intended use

* safe

REED TIP

Employers will want to know that you understand the importance of health and safety. Make sure you know the reasons for each safe working practice.

* well maintained

* used only by those who have been trained to do so.

Manual Handling Operations Regulations (1992)

These regulations try to control the risk of injury when lifting or handling bulky or heavy equipment and materials. The regulations state as follows:

* Hazardous manual handling should be avoided if possible.

* An assessment of hazardous manual handling should be made to try to find alternatives.

* You should use mechanical assistance where possible.

* The main idea is to look at how manual handling is carried out and finding safer ways of doing it.

Personal Protection at Work Regulations (PPE) (1992)

This law states that employers must provide employees with personal protective equipment (PPE) at work whenever there is a risk to health and safety. PPE needs to be:

* suitable for the work being done

* well maintained and replaced if damaged

* properly stored

* correctly used (which means employees need to be trained in how to use the PPE properly).

Work at Height Regulations (2005)

Whenever a person works at any height there is a risk that they could fall and injure themselves. The regulations place a duty on employers or anyone who controls the work of others. This means that they need to:

* plan and organise the work

* make sure those working at height are **competent**

* assess the risks and provide appropriate equipment

* manage work near or on fragile surfaces

* ensure equipment is inspected and maintained.

In all cases the regulations suggest that, if it is possible, work at height should be avoided. Perhaps the job could be done from ground level? If it is not possible, then equipment and other measures are needed to prevent the risk of falling. When working at height measures also need to be put in place to minimise the distance someone might fall.

KEY TERMS

PPE

– personal protective equipment can include gloves, goggles and hard hats.

Competent

– to be competent an organisation or individual must have:

* sufficient knowledge of the tasks to be undertaken and the risks involved

* the experience and ability to carry out their duties in relation to the project, to recognise their limitations and take appropriate action to prevent harm to those carrying out construction work, or those affected by the work.

(*Source* HSE)

Figure 1.1 Examples of personal protective equipment

Employer responsibilities under HASAWA

HASAWA states that employers with five or more staff need their own health and safety policy. Employers must assess any risks that may be involved in their workplace and then introduce controls to reduce these risks. These risk assessments need to be reviewed regularly.

Employers also need to supply personal protective equipment (PPE) to all employees when it is needed and to ensure that it is worn when required.

Specific employer responsibilities are outlined in Table 1.1.

Employee responsibilities under HASAWA

HASAWA states that all those operating in the workplace must aim to work in a safe way. For example, they must wear any PPE provided and look after their equipment. Employees should not be charged for PPE or any actions that the employer needs to take to ensure safety.

Specific employer responsibilities are outlined in Table 1.1. Table 1.2 identifies the key employee responsibilities.

KEY TERMS

Risk

– the likelihood that a person may be harmed if they are exposed to a hazard.

Hazard

– a potential source of harm, injury or ill-health.

Near miss

– any incident, accident or emergency that did not result in an injury but could have done so.

Employer responsibility	Explanation
Safe working environment	Where possible all potential risks and hazards should be eliminated.
Adequate staff training	When new employees begin a job their induction should cover health and safety. There should be ongoing training for existing employees on risks and control measures.
Health and safety information	Relevant information related to health and safety should be available for employees to read and have their own copies.
Risk assessment	Each task or job should be investigated and potential risks identified so that measures can be put in place. A risk assessment and method statement should be produced. The method statement will tell you how to carry out the task, what PPE to wear, equipment to use and the sequence of its use.
Supervision	A competent and experienced individual should always be available to help ensure that health and safety problems are avoided.

Table 1.1 Employer responsibilities under HASAWA

Employee responsibility	Explanation
Working safely	Employees should take care of themselves, only do work that they are competent to carry out and remove obvious hazards if they are seen.
Working in partnership with the employer	Co-operation is important and you should never interfere with or misuse any health and safety signs or equipment. You should always follow the site rules.
Reporting hazards, near misses and accidents correctly	Any health and safety problems should be reported and discussed, particularly a near miss or an actual accident.

Table 1.2 Employee responsibilities under HASAWA

KEY TERMS

Improvement notice

– this is issued by the HSE if a health or safety issue is found and gives the employer a time limit to make changes to improve health and safety.

Prohibition notice

– this is issued by the HSE if a health or safety issue involving the risk of serious personal injury is found and stops all work until the improvements to health and safety have been made.

Sub-contractor

– an individual or group of workers who are directly employed by the main contractor to undertake specific parts of the work.

Health and Safety Executive

The Health and Safety Executive (HSE) is responsible for health, safety and welfare. It carries out spot checks on different workplaces to make sure that the law is being followed.

HSE inspectors have access to all areas of a construction site and can also bring in the police. If they find a problem then they can issue an improvement notice. This gives the employer a limited amount of time to put things right.

In serious cases, the HSE can issue a prohibition notice. This means all work has to stop until the problem is dealt with. An employer, the employees or sub-contractors could be taken to court.

The roles and responsibilities of the HSE are outlined in Table 1.3.

Responsibility	Explanation
Enforcement	It is the HSE's responsibility to reduce work-related death, injury and ill health. It will use the law against those who put others at risk.
Legislation and advice	The HSE will use health and safety legislation to serve improvement or prohibition notices or even to prosecute those who break health and safety rules. Inspectors will provide advice either face-to-face or in writing on health and safety matters.
Inspection	The HSE will look at site conditions, standards and practices and inspect documents to make sure that businesses and individuals are complying with health and safety law.

Table 1.3 HSE roles and responsibilities

Sources of health and safety information

There is a wide variety of health and safety information. Most of it is available free of charge, while other organisations may make a charge to provide information and advice. Table 1.4 outlines the key sources of health and safety information.

Source	Types of information	Website
Health and Safety Executive (HSE)	The HSE is the primary source of work-related health and safety information. It covers all possible topics and industries.	www.hse.gov.uk
Construction Industry Training Board (CITB)	The national training organisation provides key information on legislation and site safety.	www.citb.co.uk
British Standards Institute (BSI)	Provides guidelines for risk management, PPE, fire hazards and many other health and safety-related areas.	www.bsigroup.com
Royal Society for the Prevention of Accidents (RoSPA)	Provides training, consultancy and advice on a wide range of health and safety issues that are aimed to reduce work related accidents and ill health.	www.rospa.com
Royal Society for Public Health (RSPH)	Has a range of qualifications and training programmes focusing on health and safety.	www.rsph.org.uk

Table 1.4 Health and safety information

Informing the HSE

The HSE requires the reporting of:

* deaths and injuries – any **major injury**, **over 7-day injury** or death

* occupational disease

* dangerous occurrence – a collapse, explosion, fire or collision

* gas accidents – any accidental leaks or other incident related to gas.

Enforcing guidance

Work-related injuries and illnesses affect huge numbers of people. According to the HSE, 1.1 million working people in the UK suffered from a work-related illness in 2011 to 2012. Across all industries, 173 workers were killed, 111,000 other injuries were reported and 27 million working days were lost.

The construction industry is a high risk one and, although only around 5 per cent of the working population is in construction, it accounts for 10 per cent of all major injuries and 22 per cent of fatal injuries.

The good news is that enforcing guidance on health and safety has driven down the numbers of injuries and deaths in the industry. Only 20 years ago over 120 construction workers died in workplace accidents each year. This is now reduced to fewer than 60 a year.

However, there is still more work to be done and it is vital that organisations such as the HSE continue to enforce health and safety and continue to reduce risks in the industry.

On-site safety inductions and toolbox talks

The HSE suggests that all new workers arriving on site should attend a short induction session on health and safety. It should:

* show the commitment of the company to health and safety

* explain the health and safety policy

* explain the roles individuals play in the policy

* state that each individual has a legal duty to contribute to safe working

* cover issues like excavations, work at height, electricity and fire risk

* provide a layout of the site and show evacuation routes

* identify where fire fighting equipment is located

* ensure that all employees have evidence of their skills

* stress the importance of signing in and out of the site.

Behaviour and actions that could affect others

It is the responsibility of everyone on site not only to look after their own health and safety, but also to ensure that their actions do not put anyone else at risk.

Trying to carry out work that you are not competent to do is not only dangerous to yourself but could compromise the safety of others.

Simple actions, such as ensuring that all of your rubbish and waste is properly disposed of, will go a long way to removing hazards on site that could affect others.

Just as you should not create a hazard, ignoring an obvious one is just as dangerous. You should always obey site rules and particularly the health and safety rules. You should follow any instructions you are given.

ACCIDENT AND EMERGENCY PROCEDURES

All sites will have specific procedures for dealing with accidents and emergencies. An emergency will often mean that the site needs to be evacuated, so you should know in advance where to assemble and who to report to. The site should never be re-entered without authorisation from an individual in charge or the emergency services.

Types of emergencies

Emergencies are incidents that require immediate action. They can include:

* fires
* spillages or leaks of chemicals or other hazardous substances, such as gas
* failure of a scaffold
* collapse of a wall or trench
* a health problem
* an injury
* bombs and security alerts.

Legislation and reporting accidents

RIDDOR (1995) puts a duty on employers, anyone who is self-employed, or an individual in control of the work, to report any serious workplace accidents, occupational diseases or dangerous occurrences (also known as near misses).

The report has to be made by these individuals and, if it is serious enough, the responsible person may have to fill out a RIDDOR report.

Figure 1.2 It's important that you know where your company's fire-fighting equipment is located

Injuries, diseases and dangerous occurrences

Construction sites can be dangerous places, as we have seen. The HSE maintains a list of all possible injuries, diseases and dangerous occurrences, particularly those that need to be reported.

Injuries

There are two main classifications of injuries: minor and major. A minor injury can usually be handled by a competent first aider, although it is often a good idea to refer the individual to their doctor or to the hospital. Typical minor injuries can include:

* minor cuts   * minor burns   * exposure to fumes.

Major injuries are more dangerous and will usually require the presence of an ambulance with paramedics. Major injuries can include:

* bone fracture   * concussion

* unconsciousness   * electric shock.

Diseases

There are several different diseases and health issues that have to be reported, particularly if a doctor notifies that a disease has been diagnosed. These include:

* poisoning   * infections

* skin diseases   * occupational cancer

* lung diseases   * hand/arm vibration syndrome.

Dangerous occurrences

Even if something happens that does not result in an injury, but could easily have done so, it is classed as a dangerous occurrence. It needs to be reported immediately and then followed up by an accident report form. Dangerous occurrences can include:

* accidental release of a substance that could damage health

* anything coming into contact with overhead power lines

* an electrical problem that caused a fire or explosion

* collapse or partial collapse of scaffolding over 5 m high.

Recording accidents and emergencies

The Reporting of Injuries, Diseases and Dangerous Occurrences Regulations (RIDDOR) (1995) requires employers to:

* report any relevant injuries, diseases or dangerous occurrences to the Health and Safety Executive (HSE)

* keep records of incidents in a formal and organised manner (for example, in an accident book or online database).

PRACTICAL TIP

An up-to-date list of dangerous occurrences is maintained by the Health and Safety Executive.

After an accident, you may need to complete an accident report form – either in writing or online. This form may be completed by the person who was injured or the first aider.

On the accident report form you need to note down:

* the casualty's personal details, e.g. name, address, occupation
* the name of the person filling in the report form
* the details of the accident.

In addition, the person reporting the accident will need to sign the form.

On site a trained first aider will be the first individual to try and deal with the situation. In addition to trying to save life, stop the condition from getting worse and getting help, they will also record the occurrence.

On larger sites there will be a health and safety officer, who would keep records and documentation detailing any accidents and emergencies that have taken place on site. All companies should keep such records; it may be a legal requirement for them to do so under RIDDOR and it is good practice to do so in case the HSE asks to see it.

Importance of reporting accidents and near misses

Reporting incidents is not just about complying with the law or providing information for statistics. Each time an accident or near miss takes place it means lessons can be learned and future problems avoided.

The accident or near miss can alert the business or organisation to a potential problem. They can then take steps to ensure that it does not occur in the future.

Major and minor injuries and near misses

RIDDOR defines a major injury as:

* a fracture (but not to a finger, thumb or toes)
* a dislocation
* an amputation
* a loss of sight in an eye
* a chemical or hot metal burn to the eye
* a penetrating injury to the eye
* an electric shock or electric burn leading to unconsciousness and/or requiring resuscitation
* hyperthermia, heat-induced illness or unconsciousness
* asphyxia
* exposure to a harmful substance
* inhalation of a substance
* acute illness after exposure to toxins or infected materials.

A minor injury could be considered as any occurrence that does not fall into any of the above categories.

A near miss is any incident that did not actually result in an injury but which could have caused a major injury if it had done so. Non-reportable near misses are useful to record as they can help to identify potential problems. Looking at a list of near misses might show patterns for potential risk.

Accident trends

We have already seen that the HSE maintains statistics on the number and types of construction accidents. The following are among the 2011/2012 construction statistics:

* There were 49 fatalities.

* There were 5,000 occupational cancer patients.

* There were 74,000 cases of work-related ill health.

* The most common types of injury were caused by falls, although many injuries were caused by falling objects, collapses and electricity. A number of construction workers were also hurt when they slipped or tripped, or were injured while lifting heavy objects.

Accidents, emergencies and the employer

Even less serious accidents and injuries can cost a business a great deal of money. But there are other costs too:

* Poor company image – if a business does not have health and safety controls in place then it may get a reputation for not caring about its employees. The number of accidents and injuries may be far higher than average.

* Loss of production – the injured individual might have to be treated and then may need a period of time off work to recover. The loss of production can include those who have to take time out from working to help the injured person and the time of a manager or supervisor who has to deal with all the paperwork and problems.

* Insurance – each time there is an accident or injury claim against the company's insurance the premiums will go up. If there are many accidents and injuries the business may find it impossible to get insurance. It is a legal requirement for a business to have insurance so in the end that company might have to close down.

* Closure of site – if there is a serious accident or injury then the site may have to be closed while investigations take place to discover the reason, or who was responsible. This could cause serious delays and loss of income for workers and the business.

DID YOU KNOW?

RoSPA (the Royal Society for the Prevention of Accidents) uses many of the statistics from the HSE. The latest figures that RoSPA has analysed date back to 2008/2009. In that year, 1.2 million people in the UK were suffering from work-related illnesses. With fewer than 132,000 reportable injuries at work, this is believed to be around half of the real figure.

DID YOU KNOW?

An employee working in a small business broke two bones in his arm. He could not return to proper duties for eight months. He lost out on wages while he was off sick and, in total, it cost the business over £45,000.

REED TIP

On some construction sites, you may get a Health and Safety Inspector come to look round without any notice – one more reason to always be thinking about working safely.

Accident and emergency authorised personnel

Several different groups of people could be involved in dealing with accident and emergency situations. These are listed in Table 1.5.

Authorised personnel	Role
First aiders and emergency responders	These are employees on site and in the workforce who have been trained to be the first to respond to accidents and injuries. The minimum provision of an appointed person would be someone who has had basic first aid training. The appointment of a first aider is someone who has attained a higher or specific level of training. A construction site with fewer than 5 employees needs an appointed first aider. A construction site with up to 50 employees requires a trained first aider, and for bigger sites at least one trained first aider is required for every 50 people.
Supervisors and managers	These have the responsibility of managing the site and would have to organise the response and contact emergency services if necessary. They would also ensure that records of any accidents are completed and up to date and notify the HSE if required.
Health and Safety Executive	The HSE requires businesses to investigate all accidents and emergencies. The HSE may send an inspector, or even a team, to investigate and take action if the law has been broken.
Emergency services	Calling the emergency services depends on the seriousness of the accident. Paramedics will take charge of the situation if there is a serious injury and if they feel it necessary will take the individual to hospital.

Table 1.5 People who deal with accident and emergency situations

DID YOU KNOW?

The three main emergency services in the UK are: the Fire Service (for fire and rescue); the Ambulance Service (for medical emergencies); the Police (for an immediate police response). Call them on 999 only if it is an emergency.

The basic first aid kit

BS 8599 relates to first aid kits, but it is not legally binding. The contents of a first aid box will depend on an employer's assessment of their likely needs. The HSE does not have to approve the contents of a first aid box but it states that where the work involves low level hazards the minimum contents of a first aid box should be:

* a copy of its leaflet on first aid – *HSE Basic advice on first aid at work*

* 20 sterile plasters of assorted size

* 2 sterile eye pads

* 4 sterile triangular bandages

* 6 safety pins

* 2 large sterile, unmedicated wound dressings

* 6 medium-sized sterile unmedicated wound dressings

* 1 pair of disposable gloves.

The HSE also recommends that no tablets or medicines are kept in the first aid box.

Figure 1.3 A typical first aid box

What to do if you discover an accident

When an accident happens it may not only injure the person involved directly, but it may also create a hazard that could then injure others. You need to make sure that the area is safe enough for you or someone else to help the injured person. It may be necessary to turn off the electrical supply or remove obstructions to the site of the accident.

The first thing that needs to be done if there is an accident is to raise the alarm. This could mean:

* calling for the first aider

* phoning for the emergency services

* dealing with the problem yourself.

How you respond will depend on the severity of the injury.

You should follow this procedure if you need to contact the emergency services:

* Find a telephone away from the emergency.

* Dial 999.

* You may have to go through a switchboard. Carefully listen to what the operator is saying to you and try to stay calm.

* When asked, give the operator your name and location, and the name of the emergency service or services you require.

* You will then be transferred to the appropriate emergency service, who will ask you questions about the accident and its location. Answer the questions in a clear and calm way.

* Once the call is over, make sure someone is available to help direct the emergency services to the location of the accident.

IDENTIFYING HAZARDS

As we have already seen, construction sites are potentially dangerous places. The most effective way of handling health and safety on a construction site is to spot the hazards and deal with them before they can cause an accident or an injury. This begins with basic housekeeping and carrying out risk assessments. It also means having a procedure in place to report hazards so that they can be dealt with.

Good housekeeping

Work areas should always be clean and tidy. Sites that are messy, strewn with materials, equipment, wires and other hazards can prove to be very dangerous. You should:

* always work in a tidy way
* never block fire exits or emergency escape routes
* never leave nails and screws scattered around
* ensure you clean and sweep up at the end of each working day
* not block walkways
* never overfill skips or bins
* never leave food waste on site.

Risk assessments and method statements

It is a legal requirement for employers to carry out risk assessments. This covers not only those who are actually working on a particular job, but other workers in the immediate area, and others who might be affected by the work.

It is important to remember that when you are carrying out work your actions may affect the safety of other people. It is important, therefore, to know whether there are any potential hazards. Once you know what these hazards are you can do something to either prevent or reduce them as a risk. Every job has potential hazards.

There are five simple steps to carrying out a risk assessment, which are shown in Table 1.6, using the example of repointing brickwork on the front face of a dwelling.

Step	Action	Example
1	Identify hazards	The property is on a street with a narrow pavement. The damaged brickwork and loose mortar need to be removed and placed in a skip below. Scaffolding has been erected. The road is not closed to traffic.
2	Identify who is at risk	The workers repointing are at risk as they are working at height. Pedestrians and vehicles passing are at risk from the positioning of the skip and the chance that debris could fall from height.
3	What is the risk from the hazard that may cause an accident?	The risk to the workers is relatively low as they have PPE and the scaffolding has been correctly erected. The risk to those passing by is higher, as they are unaware of the work being carried out above them.
4	Measures to be taken to reduce the risk	Station someone near the skip to direct pedestrians and vehicles away from the skip while the work is being carried out. Fix a secure barrier to the edge of the scaffolding to reduce the chance of debris falling down. Lower the bricks and mortar debris using a bucket or bag into the skip and not throwing them from the scaffolding. Consider carrying out the work when there are fewer pedestrians and less traffic on the road.
5	Monitor the risk	If there are problems with the first stages of the job, you need to take steps to solve them. If necessary consider taking the debris by hand through the building after removal.

Table 1.6 A five-step risk assessment for repointing brickwork

Your employer should follow these working practices, which can help to prevent accidents or dangerous situations occurring in the workplace:

* *Risk assessments* look carefully at what could cause an individual harm and how to prevent this. This is to ensure that no one should be injured or become ill as a result of their work. Risk assessments identify how likely it is that an accident might happen and the consequences of it happening. A risk factor is worked out and control measures created to try to offset them.

* *Method statements,* however brief, should be available for every risk assessment. They summarise risk assessments and other findings to provide guidance on how the work should be carried out.

* *Permit to work systems* are used for very high risk or even potentially fatal activities. They are checklists that need to be completed before the work begins. They must be signed by a supervisor.

* *A hazard book* lists standard tasks and identifies common hazards. These are useful tools to help quickly identify hazards related to particular tasks.

Types of hazards

Typical construction accidents can include:

* fires and explosions
* slips, trips and falls.
* burns, including those from chemicals
* falls from scaffolding, ladders and roofs
* electrocution
* injury from faulty machinery
* power tool accidents
* being hit by construction debris
* falling through holes in flooring

We will look at some of the more common hazards in a little more detail.

Fires

Fires need oxygen, heat and fuel to burn. Even a spark can provide enough heat needed to start a fire, and anything flammable, such as petrol, paper or wood, provides the fuel. It may help to remember the 'triangle of fire' – heat, oxygen and fuel are all needed to make fire so remove one or more to help prevent or stop the fire.

Tripping

Leaving equipment and materials lying around can cause accidents, as can trailing cables and spilt water or oil. Some of these materials are also potential fire hazards.

Chemical spills

If the chemicals are not hazardous then they just need to be mopped up. But sometimes they do involve hazardous materials and there will be an existing plan on how to deal with them. A risk assessment will have been carried out.

Falls from height

A fall even from a low height can cause serious injuries. Precautions need to be taken when working at height to avoid permanent injury. You should also consider falls into open excavations as falls from height. All the same precautions need to be in place to prevent a fall.

Burns

Burns can be caused not only by fires and heat, but also from chemicals and solvents. Electricity and wet concrete and cement can also burn skin. PPE is often the best way to avoid these dangers. Sunburn is a common and uncomfortable form of burning and sunscreen should be made available. For example keeping skin covered up will help to prevent sunburn. You might think a tan looks good, but it could lead to skin cancer.

Electrical

Electricity is hazardous and electric shocks can cause burns and muscle damage, and can kill.

Exposure to hazardous substances

We look at hazardous substances in more detail on pages 20–1. COSHH regulations identify hazardous substances and require them to be labelled. You should always follow the instructions when using them.

Plant and vehicles

On busy sites there is always a danger from moving vehicles and heavy plant. Although many are fitted with reversing alarms, it may not be easy to hear them over other machinery and equipment. You should always ensure you are not blocking routes or exits. Designated walkways separate site traffic and pedestrians – this includes workers who are walking around the site. Crossing points should be in place for ease of movement on site.

Reporting hazards

We have already seen that hazards have the potential to cause serious accidents and injuries. It is therefore important to report hazards and there are different methods of doing this.

The first major reason to report hazards is to prevent danger to others, whether they are other employees or visitors to the site. It is vital to prevent accidents from taking place and to quickly correct any dangerous situations.

Injuries, diseases and actual accidents all need to be reported and so do dangerous occurrences. These are incidents that do not result in an actual injury, but could easily have hurt someone.

Accidents need to be recorded in an accident book, computer database or other secure recording system, as do near misses. Again it is a legal requirement to keep appropriate records of accidents and every company will have a procedure for this which they should tell you about. Everyone should know where the book is kept or how the records are made. Anyone that has been hurt or has taken part in dealing with an occurrence should complete the details of what has happened. Typically this will require you to fill in:

* the date, time and place of the incident

* how it happened

* what was the cause

* how it was dealt with

* who was involved

* signature and date.

The details in the book have to be transferred onto an official HSE report form.

As far as is possible, the site, company or workplace will have set procedures in place for reporting hazards and accidents. These procedures will usually be found in the place where the accident book or records are stored. The location tends to be posted on the site notice board.

How hazards are created

Construction sites are busy places. There are constantly new stages in development. As each stage is begun a whole new set of potential hazards need to be considered.

At the same time, new workers will always be joining the site. It is mandatory for them to be given health and safety instruction during induction. But sometimes this is impossible due to pressure of work or availability of trainers.

Construction sites can become even more hazardous in times of extreme weather:

* Flooding – long periods of rain can cause trenches to fill with water, cellars to be flooded and smooth surfaces to become extremely wet and slippery.

* Wind – strong winds may prevent all work at height. Scaffolding may have become unstable, unsecured roofing materials may come loose, dry-stored materials such as sand and cement may have been blown across the site.

* Heat – this can change the behaviour of materials: setting quicker, failing to cure and melting. It can also seriously affect the health of the workforce through dehydration and heat exhaustion.

* Snow – this can add enormous weight to roofs and other structures and could cause collapse. Snow can also prevent access or block exits and can mean that simple and routine work becomes impossible due to frozen conditions.

Storing combustibles and chemicals

A combustible substance can be both flammable and explosive. There are some basic suggestions from the HSE about storing these:

* Ventilation – the area should be well ventilated to disperse any vapours that could trigger off an explosion.

* Ignition – an ignition is any spark or flame that could trigger off the vapours, so materials should be stored away from any area that uses electrical equipment or any tool that heats up.

* Containment – the materials should always be kept in proper containers with lids and there should be spillage trays to prevent any leak seeping into other parts of the site.

* Exchange – in many cases it can be possible to find an alternative material that is less dangerous. This option should be taken if possible.

* Separation – always keep flammable substances away from general work areas. If possible they should be partitioned off.

Combustible materials can include a large number of commonly used substances, such as cleaning agents, paints and adhesives.

HEALTH AND HYGIENE

Just as hazards can be a major problem on site, other less obvious problems relating to health and hygiene can also be an issue. It is both your responsibility and that of your employer to make sure that you stay healthy.

The employer will need to provide basic welfare facilities, no matter where you are working and these must have minimum standards.

Welfare facilities

Welfare facilities can include a wide range of different considerations, as can be seen in Table 1.7.

Facilities	Purpose and minimum standards
Toilets	If there is a lock on the door there is no need to have separate male and female toilets. There should be enough for the site workforce. If there is no flushing water on site they must be chemical toilets.
Washing facilities	There should be a wash basin large enough to be able to wash up to the elbow. There should be soap, hot and cold water and, if you are working with dangerous substances, then showers are needed.
Drinking water	Clean drinking water should be available; either directly connected to the mains or bottled water. Employers must ensure that there is no contamination.
Dry room	This can operate also as a store room, which needs to be secure so that workers can leave their belongings there and also use it as a place to dry out if they have been working in wet weather, in which case a heater needs to be provided.
Work break area	This is a shelter out of the wind and rain, with a kettle, a microwave, tables and chairs. It should also have heating.

Table 1.7 Welfare facilities in the workplace

CASE STUDY

South
Tyneside Homes

South Tyneside Council's
Housing Company

Staying safe on site

Johnny McErlane finished his apprenticeship at South Tyneside Homes a year ago.

'I've been working on sheltered accommodation for the last year, so there are a lot of vulnerable and elderly people around. All the things I learnt at college from doing the health and safety exams comes into practice really, like taking care when using extension leads, wearing high-vis and correct footwear. It's not just about your health and safety, but looking out for others as well.

On the shelters, you can get a health and safety inspector who just comes around randomly, so you have to always be ready. It just becomes a habit once it's been drilled into you. You're health and safety conscious all the time.

The shelters also have a fire alarm drill every second Monday, so you've got to know the procedure involved there. When it comes to the more specialised skills, such as mouth-to-mouth and CPR, you might have a designated first aider on site who will have their skills refreshed regularly. Having a full first aid certificate would be valuable if you're working in construction.

You cover quite a bit of the first aid skills in college and you really have to know them because you're not always working on large sites. For example, you might be on the repairs team, working in people's houses where you wouldn't have a first aider, so you've got to have the basic knowledge yourself, just in case. All our vans have a basic first aid kit that's kept fully stocked.

The company keeps our knowledge current with these "toolbox talks", which are like refresher courses. They give you any new information that needs to be passed on to all the trades. It's a good way of keeping everyone up to date.'

Noise

Ear defenders are the best precaution to protect the ears from loud noises on site. Ear defenders are either basic ear plugs or ear muffs, which can be seen in Fig 1.13 on page 32.

The long-term impact of noise depends on the intensity and duration of the noise. Basically, the louder and longer the noise exposure, the more damage is caused. There are ways of dealing with this:

* Remove the source of the noise.

* Move the equipment away from those not directly working with it.

* Put the source of the noise into a soundproof area or cover it with soundproof material.

* Ask a supervisor if they can move all other employees away from that part of the site until the noise stops.

Substances hazardous to health

COSHH Regulations (see page 3) identify a wide variety of substances and materials that must be labelled in different ways.

Controlling the use of these substances is always difficult. Ideally, their use should be eliminated (stopped) or they should be replaced with something less harmful. Failing this, they should only be used in controlled or restricted areas. If none of this is possible then they should only be used in controlled situations.

If a hazardous situation occurs at work, then you should:

* ensure the area is made safe

* inform the supervisor, site manager, safety officer or other nominated person.

You will also need to report any potential hazards or near misses.

Personal hygiene

Construction sites can be dirty places to work. Some jobs will expose you to dust, chemicals or substances that can make contact with your skin or may stain your work clothing. It is good practice to wear suitable PPE as a first line of defence as chemicals can penetrate your skin. Whenever you have finished a job you should always wash your hands. This is certainly true before eating lunch or travelling home. It can be good practice to have dedicated work clothing, which should be washed regularly.

Always ensure you wash your hands and face and scrub your nails. This will prevent dirt, chemicals and other substances from contaminating your food and your home.

Make sure that you regularly wash your work clothing and either repair it or replace it if it becomes too worn or stained.

Health risks

The construction industry uses a wide variety of substances that could harm your health. You will also be carrying out work that could be a health risk to you, and you should always be aware that certain activities could cause long-term damage or even kill you if things go wrong. Unfortunately not all health risks are immediately obvious. It is important to make sure that from time to time you have health checks, particularly if you have been using hazardous substances. Table 1.8 outlines some potential health risks in a typical construction site.

KEY TERMS

Dermatitis

– this is an inflammation of the skin. The skin will become red and sore, particularly if you scratch the area. A GP should be consulted.

Leptospirosis

– this is also known as Weil's disease. It is spread by touching soil or water contaminated with the urine of wild animals infected with the leptospira bacteria. Symptoms are usually flu-like but in extreme cases it can cause organ failure.

Health risk	Potential future problems
Dust	The most dangerous potential dust is, of course, asbestos, which **should only be handled by specialists under controlled conditions**. But even brick dust and other fine particles can cause eye injuries, problems with breathing and even cancer.
Chemicals	Inhaling or swallowing dangerous chemicals could cause immediate, long-term damage to lungs and other internal organs. Skin problems include burns or skin can become very inflamed and sore. This is known as dermatitis.
Bacteria	Contact with waste water or soil could lead to a bacterial infection. The germs in the water or dirt could cause infection which will require treatment if they enter the body. The most extreme version is leptospirosis.
Heavy objects	Lifting heavy, bulky or awkward objects can lead to permanent back injuries that could require surgery. Heavy objects can also damage the muscles in all areas of the body.
Noise	Failure to wear ear defenders when you are exposed to loud noises can permanently affect your hearing. This could lead to deafness in the future.
Vibrating tools	Using machines that vibrate can cause a condition known as hand/arm vibration syndrome (HAVS) or vibration white finger, which is caused by injury to nerves and blood vessels. You will feel tingling that could lead to permanent numbness in the fingers and hands, as well as muscle weakness.
Cuts	Any open wound, no matter how small, leaves your body exposed to potential infections. Cuts should always be cleaned and covered, preferably with a waterproof dressing. The blood loss from deep cuts could make you feel faint and weak, which may be dangerous if you are working at height or operating machinery.
Sunlight	Most construction work involves working outside. There is a temptation to take advantage of hot weather and get a tan. But long-term exposure to sunshine means risking skin cancer so you should cover up and apply sun cream.
Head injuries	You should seek medical attention after any bump to the head. Severe head injuries could cause epilepsy, hearing problems, brain damage or death.

Table 1.8 Health risks in construction

HANDLING AND STORING MATERIALS AND EQUIPMENT

On a busy construction site it is often tempting not to even think about the potential dangers of handling equipment and materials. If something needs to be moved or collected you will just pick it up without any thought. It is also tempting just to drop your tools and other equipment when you have finished with them to deal with later. But abandoned equipment and tools can cause hazards both for you and for other people.

Safe lifting

Lifting or handling heavy or bulky items is a major cause of injuries on construction sites. So whenever you are dealing with a heavy load, it is important to carry out a basic risk assessment.

The first thing you need to do is to think about the job to be done and ask:

* Do I need to lift it manually or is there another way of getting the object to where I need it?

Consider any mechanical methods of transporting loads or picking up materials. If there really is no alternative, then ask yourself:

1. Do I need to bend or twist?

2. Does the object need to be lifted or put down from high up?

3. Does the object need to be carried a long way?

4. Does the object need to be pushed or pulled for a long distance?

5. Is the object likely to shift around while it is being moved?

If the answer to any of these questions is 'yes', you may need to adjust the way the task is done to make it safer.

Think about the object itself. Ask:

1. Is it just heavy or is it also bulky and an awkward shape?

2. How easy is it to get a good hand-hold on the object?

3. Is the object a single item or are there parts that might move around and shift the weight?

4. Is the object hot or does it have sharp edges?

Again, if you have answered 'yes' to any of these questions, then you need to take steps to address these issues.

It is also important to think about the working environment and where the lifting and carrying is taking place. Ask yourself:

1. Are the floors stable?

2. Are the surfaces slippery?

3. Will a lack of space restrict my movement?

4. Are there any steps or slopes?

5. What is the lighting like?

Before lifting and moving an object, think about the following:

* Check that your pathway is clear to where the load needs to be taken.

* Look at the product data sheet and assess the weight. If you think the object is too heavy or difficult to move then ask someone to help you. Alternatively, you may need to use a mechanical lifting device.

When you are ready to lift, gently raise the load. Take care to ensure the correct posture – you should have a straight back, with your elbows tucked in, your knees bent and your feet slightly apart.

Once you have picked up the load, move slowly towards your destination. When you get there, make sure that you do not drop the load but carefully place it down.

DID YOU KNOW?

Although many people regard the weight limit for lifting and/or moving heavy or awkward objects to be 20 kg, the HSE does not recommend safe weights. There are many things that will affect the ability of an individual to lift and carry particular objects and the risk that this creates, so manual handling should be avoided altogether where possible.

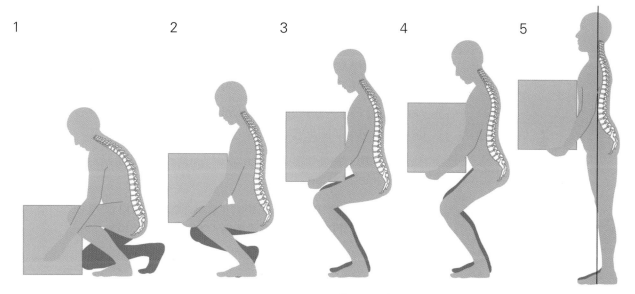

Figure 1.4 Take care to follow the correct procedure for lifting

Sack trolleys are useful for moving heavy and bulky items around. Gently slide the bottom of the sack trolley under the object and then raise the trolley to an angle of 45° before moving off. Make sure that the object is properly balanced and is not too big for the trolley.

Trailers and forklift trucks are often used on large construction sites, as are dump trucks. Never use these without proper training.

Figure 1.5 Pallet truck

Figure 1.6 Sack trolley

Site safety equipment

You should always read the construction site safety rules and when required wear your PPE. Simple things, such as wearing the right footwear for the right job, are important.

Safety equipment falls into two main categories:

* PPE – including hard hats, footwear, gloves, glasses and safety vests

* perimeter safety – this includes screens, netting and guards or clamps to prevent materials from falling or spreading.

Construction safety is also directed by signs, which will highlight potential hazards.

Safe handling of materials and equipment

All tools and equipment are potentially dangerous. It is up to you to make sure that they do not cause harm to yourself or others. You should always know how to use tools and equipment. This means either instruction from someone else who is experienced, or at least reading the manufacturer's instructions.

You should always make sure that you:

* use the right tool – don't be tempted to use a tool that is close to hand instead of the one that is right for the job

* wear your PPE – the one time you decide not to bother could be the time that you injure yourself

* never try to use a tool or a piece of equipment that you have not been trained to use.

You should always remember that if you are working on a building that was constructed before 2000 it may contain asbestos.

Correct storage

We have already seen that tools and equipment need to be treated with respect. Damaged tools and equipment are not only less effective at doing their job, they could also cause you to injure yourself.

Table 1.9 provides some pointers on how to store and handle different types of materials and equipment.

Materials and equipment	Safe storage and handling
Hand tools	Store hand tools with sharp edges either in a cover or a roll. They should be stored in bags or boxes. They should always be dried before putting them away as they will rust.
Power tools	Never carry them by the cable. Store them in their original carrying case. Always follow the manufacturer's instructions.
Wheelbarrows	Check the tyres and metal stays regularly. Always clean out after use and never overload.
Bricks and blocks	Never store more than two packs high. When cutting open a pack, be careful as the bricks could collapse.
Slabs and curbs	Store slabs flat on their edges on level ground, preferably with wood underneath to prevent damage. Store curbs the same way. To prevent weather damage, cover them with a sheet.
Tiles	Always cover them and protect them from damage as they are relatively fragile. Ideally store them in a hut or container.
Aggregates	Never store aggregates under trees as leaves will drop on them and contaminate them. Cover them with plastic sheets.
Plaster and plasterboard	Plaster needs to be kept dry, so even if stored inside you should take the precaution of putting the bags on pallets. To prevent moisture do not store against walls and do not pile higher than five bags. Plasterboard can be awkward to manage and move around. It also needs to be stored in a waterproof area. It should be stored flat and off the ground but should not be stored against walls as it may bend. Use a rotation system so that the materials are not stored in the same place for long periods.
Wood	Always keep wood in dry, well-ventilated conditions. If it needs to be stored outside it should be stored on bearers that may be on concrete. If wood gets wet and bends it is virtually useless. Always be careful when moving large cuts of wood or sheets of ply or MDF as they can easily become damaged.
Adhesives and paint	Always read the manufacturer's instructions. Ideally they should always be stored on clearly marked shelves. Make sure you rotate the stock using the older stock first. Always make sure that containers are tightly sealed. Storage areas must comply with fire regulations and display signs to advise of their contents.

Table 1.9 Safe storing and handling of materials and equipment

Waste control

The expectation within the building services industry is increasingly that working practices conserve energy and protect the environment. Everyone can play a part in this. For example, you can contribute by turning off hose pipes when you have finished using water, or not running electrical items when you don't need to.

Simple things, such as keeping construction sites neat and orderly, can go a long way to conserving energy and protecting the environment. A good way to remember this is Sort, Set, Shine, Standardise:

* Sort – sort and store items in your work area, eliminate clutter and manage deliveries.

* Set – everything should have its own place and be clearly marked and easy to access. In other words, be neat!

Figure 1.7 It's important to create as little waste as possible on the construction site

* Shine – clean your work area and you will be able to see potential problems far more easily.

* Standardise – by using standardised working practices you can keep organised, clean and safe.

Reducing waste is all about good working practice. By reducing wastage disposal, and recycling materials on site, you will benefit from savings on raw materials and lower transportation costs.

Planning ahead, and accurately measuring and cutting materials, means that you will be able to reduce wastage.

BASIC WORKING PLATFORMS AND ACCESS EQUIPMENT

Working at height should be eliminated or the work carried out using other methods where possible. However, there may be situations where you may need to work at height. These situations can include:

* roofing

* repair and maintenance above ground level

* working on high ceilings.

Any work at height must be carefully planned. Access equipment includes all types of ladder, scaffold and platform. You must always use a working platform that is safe. Sometimes a simple step ladder will be sufficient, but at other times you may have to use a tower scaffold.

Generally, ladders are fine for small, quick jobs of less than 30 minutes. However, for larger, longer jobs a more permanent piece of access equipment will be necessary.

Working platforms and access equipment: good practice and dangers of working at height

Table 1.10 outlines the common types of equipment used to allow you to work at heights, along with the basic safety checks necessary.

Equipment	Main features	Safety checks
Step ladder	Ideal for confined spaces. Four legs give stability	• Knee should remain below top of steps • Check hinges, cords or ropes • Position only to face work
Ladder	Ideal for basic access, short-term work. Made from aluminium, fibreglass or wood	• Check rungs, tie rods, repairs, and ropes and cords on stepladders • Ensure it is placed on firm, level ground • Angle should be no greater than 75° or 1 in 4
Mobile mini towers or scaffolds	These are usually aluminium and foldable, with lockable wheels	• Ensure the ground is even and the wheels are locked • Never move the platform while it has tools, equipment or people on it
Roof ladders and crawling boards	The roof ladder allows access while crawling boards provide a safe passage over tiles	• The ladder needs to be long enough and supported • Check boards are in good condition • Check the welds are intact • Ensure all clips function correctly
Mobile tower scaffolds	These larger versions of mini towers usually have edge protection	• Ensure the ground is even and the wheels are locked • Never move the platform while it has tools, equipment or people on it • Base width to height ratio should be no greater than 1:3
Fixed scaffolds and edge protection	Scaffolds fitted and sized to the specific job, with edge protection and guard rails	• There needs to be sufficient braces, guard rails and scaffold boards • The tubes should be level • There should be proper access using a ladder
Mobile elevated work platforms	Known as scissor lifts or cherry pickers	• Specialist training is required before use • Use guard rails and toe boards • Care needs to be taken to avoid overhead hazards such as cables

Table 1.10 Equipment for working at height and safety checks

You must be trained in the use of certain types of access equipment, like mobile scaffolds. Care needs to be taken when assembling and using access equipment. These are all examples of good practice:

* Step ladders should always rest firmly on the ground. Only use the top step if the ladder is part of a platform.

* Do not rest ladders against fragile surfaces, and always use both hands to climb. It is best if the ladder is steadied (footed) by someone at the foot of the ladder. Always maintain three points of contact – two feet and one hand.

* A roof ladder is positioned by turning it on its wheels and pushing it up the roof. It then hooks over the ridge tiles. Ensure that the access ladder to the roof is directly beside the roof ladder.

* A mobile scaffold is put together by slotting sections until the required height is reached. The working platform needs to have a suitable edge protection such as guard-rails and toe-boards. Always push from the bottom of the base and not from the top to move it, otherwise it may lean or topple over.

Figure 1.8 A tower scaffold

WORKING SAFELY WITH ELECTRICITY

It is essential whenever you work with electricity that you are competent and that you understand the common dangers. Electrical tools must be used in a safe manner on site. There are precautions that you can take to prevent possible injury, or even death.

Precautions

Whether you are using electrical tools or equipment on site, you should always remember the following:

* Use the right tool for the job.

* Use a transformer with equipment that runs on 110V.

* Keep the two voltages separate from each other. You should avoid using 230V where possible but, if you must, use a residual current device (RCD) if you have to use 230V.

* When using 110V, ensure that leads are yellow in colour.

* Check the plug is in good order

* Confirm that the fuse is the correct rating for the equipment.

* Check the cable (including making sure that it does not present a tripping hazard).

* Find out where the mains switch is, in case you need to turn off the power in the event of an emergency.

* Never attempt to repair electrical equipment yourself.

* Disconnect from the mains power before making adjustments, such as changing a drill bit.

* Make sure that the electrical equipment has a sticker that displays a recent test date.

Visual inspection and testing is a three-stage process:

1. The user should check for potential danger signs, such as a frayed cable or cracked plug.

2. A formal visual inspection should then take place. If this is done correctly then most faults can be detected.

3. Combined inspections and **PAT** should take place at regular intervals by a competent person.

Watch out for the following causes of accidents – they would also fail a safety check:

KEY TERMS

PAT

– Portable Appliance Testing – regular testing is a health and safety requirement under the Electricity at Work Regulations (1989).

* damage to the power cable or plug

* taped joints on the cable

* wet or rusty tools and equipment

* weak external casing

* loose parts or screws

* signs of overheating

* the incorrect fuse

* lack of cord grip

* electrical wires attached to incorrect terminals

* bare wires.

When preparing to work on an electrical circuit, do not start until a permit to work has been issued by a supervisor or manager to a competent person.

Make sure the circuit is broken before you begin. A 'dead' circuit will not cause you, or anybody else, harm. These steps must be followed:

* Switch off – ensure the supply to the circuit is switched off by disconnecting the supply cables or using an isolating switch.

* Isolate – disconnect the power cables or use an isolating switch.

* Warn others – to avoid someone reconnecting the circuit, place warning signs at the isolation point.

* Lock off – this step physically prevents others from reconnecting the circuit.

* Testing – is carried out by electricians but you should be aware that it involves three parts:

  1. testing a voltmeter on a known good source (a live circuit) so you know it is working properly

  2. checking that the circuit to be worked on is dead

  3. rechecking your voltmeter on the known live source, to prove that it is still working properly.

It is important to make sure that the correct point of isolation is identified. Isolation can be next to a local isolation device, such as a plug or socket, or a circuit breaker or fuse.

The isolation should be locked off using a unique key or combination. This will prevent access to a main isolator until the work has been completed. Alternatively, the handle can be made detachable in the OFF position so that it can be physically removed once the circuit is switched off.

Dangers

You are likely to encounter a number of potential dangers when working with electricity on construction sites or in private houses. Table 1.11 outlines the most common dangers.

Danger	Identifying the danger
Faulty electrical equipment	Visually inspect for signs of damage. Equipment should be double insulated or incorporate an earth cable.
Damaged or worn cables	Check for signs of wear or damage regularly. This includes checking power tools and any wiring in the property.
Trailing cables	Cables lying on the ground, or worse, stretched too far, can present a tripping hazard. They could also be cut or damaged easily.
Cables and pipe work	Always treat services you find as though they are live. This is very important as services can be mistaken for one another. You may have been trained to use a cable and pipe locator that finds cables and metal pipes.
Buried or hidden cables	Make sure you have plans. Alternatively, use a cable and pipe locator, mark the positions, look out for signs of service connection cables or pipes and hand-dig trial holes to confirm positions.
Inadequate over-current protection	Check circuit breakers and fuses are the correct size current rating for the circuit. A qualified electrician may have to identify and label these.

Table 1.11 Common dangers when working with electricity

Each year there are around 1,000 accidents at work involving electric shocks or burns from electricity. If you are working in a construction site you are part of a group that is most at risk. Electrical accidents happen when you are working close to equipment that you think is disconnected but which is, in fact, live.

Another major danger is when electrical equipment is either misused or is faulty. Electricity can cause fires and contact with the live parts can give you an electric shock or burn you.

Different voltages

The two most common voltages that are used in the UK are 230V and 110V:

* 230V: this is the standard domestic voltage. But on construction sites it is considered to be unsafe and therefore 110V is commonly used.

* 110V: these plugs are marked with a yellow casement and they have a different shaped plug. A transformer is required to convert 230V to 110V.

Some larger homes, as well as industrial and commercial buildings, may have 415V supplies. This is the same voltage that is found on overhead electricity cables. In most houses and other buildings the voltage from these cables is reduced to 230V. This is what most electrical equipment works from. Some larger machinery actually needs 415V.

In these buildings the 415V comes into the building and then can either be used directly or it is reduced so that normal 230V appliances can be used.

Colour coded cables

Normally you will come across three differently coloured wires: Live, Neutral and Earth. These have standard colours that comply with European safety standards and to ensure that they are easily identifiable. However, in some older buildings the colours are different.

Wire type	Modern colour	Older colour
Live	Brown	Red
Neutral	Blue	Black
Earth	Yellow and Green	Yellow and Green

Table 1.12 Colour coding of cables

Working with equipment with different electrical voltages

You should always check that the electrical equipment that you are going to use is suitable for the available electrical supply. The equipment's power requirements are shown on its rating plate. The voltage from the supply needs to match the voltage that is required by the equipment.

Storing electrical equipment

Electrical equipment should be stored in dry and secure conditions. Electrical equipment should never get wet but – if it does happen – it should be dried before storage. You should always clean and adjust the equipment before connecting it to the electricity supply.

PERSONAL PROTECTIVE EQUIPMENT (PPE)

Personal protective equipment, or PPE, is a general term that is used to describe a variety of different types of clothing and equipment that aim to help protect against injuries or accidents. Some PPE you will use on a daily basis and others you may use from time to time. The type of PPE you wear depends on what you are doing and where you are. For example, the practical exercises in this book were photographed at a college, which has rules and requirements for PPE that are different to those on large construction sites. Follow your tutor's or employer's instructions at all times.

Types of PPE

PPE literally covers from head to foot. Here are the main PPE types.

Figure 1.9 A hi-vis jacket

Figure 1.10 Safety glasses and goggles

Figure 1.11 Hand protection

Figure 1.12 Head protection

Figure 1.13 Hearing protection

Protective clothing

Clothing protection such as overalls:

* provides some protection from spills, dust and irritants
* can help protect you from minor cuts and abrasions
* reduces wear to work clothing underneath.

Sometimes you may need waterproof or chemical-resistant overalls.

High visibility (hi-vis) clothing stands out against any background or in any weather conditions. It is important to wear high visibility clothing on a construction site to ensure that people can see you easily. In addition, workers should always try to wear light-coloured clothing underneath, as it is easier to see.

You need to keep your high visibility and protective clothing clean and in good condition.

Employers need to make sure that employees understand the reasons for wearing high visibility clothing and the consequences of not doing so.

Eye protection

For many jobs, it is essential to wear goggles or safety glasses to prevent small objects, such as dust, wood or metal, from getting into the eyes. As goggles tend to steam up, particularly if they are being worn with a mask, safety glasses can often be a good alternative.

Hand protection

Wearing gloves will help to prevent damage or injury to the hands or fingers. For example, general purpose gloves can prevent cuts, and rubber gloves can prevent skin irritation and inflammation, such as contact dermatitis caused by handling hazardous substances. There are many different types of gloves available, including specialist gloves for working with chemicals.

Head protection

Hard hats or safety helmets are compulsory on building sites. They can protect you from falling objects or banging your head. They need to fit well and they should be regularly inspected and checked for cracks. Worn straps mean that the helmet should be replaced, as a blow to the head can be fatal. Hard hats bear a date of manufacture and should be replaced after about 3 years.

Hearing protection

Ear defenders, such as ear protectors or plugs, aim to prevent damage to your hearing or hearing loss when you are working with loud tools or are involved in a very noisy job.

Respiratory protection

Breathing in fibre, dust or some gases could damage the lungs. Dust is a very common danger, so a dust mask, face mask or respirator may be necessary.

Make sure you have the right mask for the job. It needs to fit properly otherwise it will not give you sufficient protection.

Foot protection

Foot protection is compulsory on site. Footwear should include steel toecaps (or equivalent) to protect feet from dropped objects, midsole protection (usually a steel plate) to protect against puncture or penetration from things like nails on the floor, and soles with good grip to help prevent slips on wet surfaces.

Figure 1.14 Respiratory protection

Legislation covering PPE

The most important piece of legislation is the Personal Protective Equipment at Work Regulations (1992). It covers all sorts of PPE and sets out your responsibilities and those of the employer. Linked to this are the Control of Substances Hazardous to Health (2002) and the Provision and Use of Work Equipment Regulations (1992 and 1998).

Storing and maintaining PPE

All forms of PPE will be less effective if they are not properly maintained. This may mean examining the PPE and either replacing or cleaning it, or if relevant testing or repairing it. PPE needs to be stored properly so that it is not damaged, contaminated or lost. Each type of PPE should have a CE mark. This shows that it has met the necessary safety requirements.

Importance of PPE

PPE needs to be suitable for its intended use and it needs to be used in the correct way. As a worker or an employee you need to:

* make sure you are trained to use PPE

* follow your employer's instructions when using the PPE and always wear it when you are told to do so

* look after the PPE and if there is a problem with it report it.

Your employer will:

* know the risks that the PPE will either reduce or avoid

* know how the PPE should be maintained

* know its limitations.

Consequences of not using PPE

The consequences of not using PPE can be immediate or long-term. Immediate problems are more obvious, as you may injure yourself. The longer-term consequences could be ill health in the future. If your employer has provided PPE, you have a legal responsibility to wear it.

FIRE AND EMERGENCY PROCEDURES

KEY TERMS

Assembly point

– an agreed place outside the building to go to if there is an emergency.

If there is a fire or an emergency, it is vital that you raise the alarm quickly. You should leave the building or site and then head for the **assembly point.**

When there is an emergency a general alarm should sound. If you are working on a larger and more complex construction site, evacuation may begin by evacuating the area closest to the emergency. Areas will then be evacuated one-by-one to avoid congestion of the escape routes.

Figure 1.15 Assembly point sign

Three elements essential to creating a fire

Three ingredients are needed to make something combust (burn):

* oxygen   * heat   * fuel.

The fuel can be anything which burns, such as wood, paper or flammable liquids or gases, and oxygen is in the air around us, so all that is needed is sufficient heat to start a fire.

The fire triangle represents these three elements visually. By removing one of the three elements the fire can be prevented or extinguished.

Figure 1.16 The fire triangle

How fire is spread

Fire can easily move from one area to another by finding more fuel. You need to consider this when you are storing or using materials on site, and be aware that untidiness can be a fire risk. For example, if there are wood shavings on the ground the fire can move across them, burning up the shavings.

Heat can also transfer from one source of fuel to another. If a piece of wood is on fire and is against or close to another piece of wood, that too will catch fire and the fire will have spread.

On site, fires are classified according to the type of material that is on fire. This will determine the type of fire-fighting equipment you will need to use. The five different types of fire are shown in Table 1.13.

Class of fire	Fuel or material on fire
A	Wood, paper and textiles
B	Petrol, oil and other flammable liquids
C	LPG, propane and other flammable gases
D	Metals and metal powder
E	Electrical equipment

Table 1.13 Different classes of fire

There is also F, cooking oil, but this is less likely to be found on site, except in a kitchen.

Taking action if you discover a fire and fire evacuation procedures

During induction, you will have been shown what to do in the event of a fire and told about assembly points. These are marked by signs and somewhere on the site there will be a map showing their location.

If you discover a fire you should:

* sound the alarm

* not attempt to fight the fire unless you have had fire marshal training

* otherwise stop work, do not collect your belongings, do not run, and do not re-enter the site until the all clear has been given.

Different types of fire extinguishers

Extinguishers can be effective when tackling small localised fires. However, you must use the correct type of extinguisher. For example, putting water on an oil fire could make it explode. For this reason, you should not attempt to use a fire extinguisher unless you have had proper training.

When using an extinguisher it is important to remember the following safety points:

* Only use an extinguisher at the early stages of a fire, when it is small.

* The instructions for use appear on the extinguisher.

* If you do choose to fight the fire because it is small enough, and you are sure you know what is burning, position yourself between the fire and the exit, so that if it doesn't work you can still get out.

Type of fire risk	Fire class Symbol	White label Water	Cream label Foam	Black label Carbon dioxide	Blue label Dry powder	Yellow label Wet chemical
A – Solid (e.g. wood or paper)	A	✓	✓	✗	✓	✓
B – Liquid (e.g. petrol)	B	✗	✓	✓	✓	✗
C – Gas (e.g. propane)	C	✗	✗	✓	✓	✗
D – Metal (e.g. aluminium)	D METAL	✗	✗	✗	✓	✗
E – Electrical (i.e. any electrical equipment)	E	✗	✗	✓	✓	✗
F – Cooking oil (e.g. a chip pan)	F	✗	✗	✗	✗	✓

Table 1.14 Types of fire extinguishers

There are some differences you should be aware of when using different types of extinguisher:

- *CO₂ extinguishers* – do not touch the nozzle; simply operate by holding the handle. This is because the nozzle gets extremely cold when ejecting the CO_2, as does the canister. Fires put out with a CO_2 extinguisher may reignite, and you will need to ventilate the room after use.

- *Powder extinguishers* – these can be used on lots of kinds of fire, but can seriously reduce visibility by throwing powder into the air as well as on the fire.

SIGNS AND SAFETY NOTICES

In a well-organised working environment safety signs will warn you of potential dangers and tell you what to do to stay safe. They are used to warn you of hazards. Their purpose is to prevent accidents. Some will tell you what to do (or not to do) in particular parts of the site and some will show you where things are, such as the location of a first aid box or a fire exit.

Types of signs and safety notices

There are five basic types of safety sign, as well as signs that are a combination of two or more of these types. These are shown in Table 1.15.

Type of safety sign	What it tells you	What it looks like	Example
Prohibition sign	Tells you what you must *not* do	Usually round, in red and white	Do not use ladder
Hazard sign	Warns you about hazards	Triangular, in yellow and black	Caution Slippery floor
Mandatory sign	Tells you what you *must* do	Round, usually blue and white	Masks must be worn in this area
Safe condition or information sign	Gives important information, e.g. about where to find fire exits, assembly points or first aid kit, or about safe working practices	Green and white	First aid
Firefighting sign	Gives information about extinguishers, hydrants, hoses and fire alarm call points, etc.	Red with white lettering	Fire alarm call point
Combination sign	These have two or more of the elements of the other types of sign, e.g. hazard, prohibition and mandatory		DANGER Isolate before removing cover

Table 1.15 Different types of safety signs

TEST YOURSELF

1. Which of the following requires you to tell the HSE about any injuries or diseases?

   a. HASAWA

   b. COSHH

   c. RIDDOR

   d. PUWER

2. What is a prohibition notice?

   a. An instruction from the HSE to stop all work until a problem is dealt with

   b. A manufacturer's announcement to stop all work using faulty equipment

   c. A site contractor's decision not to use particular materials

   d. A local authority banning the use of a particular type of brick

3. Which of the following is considered a major injury?

   a. Bruising on the knee

   b. Cut

   c. Concussion

   d. Exposure to fumes

4. If there is an accident on a site who is likely to be the first to respond?

   a. First aider

   b. Police

   c. Paramedics

   d. HSE

5. Which of the following is a summary of risk assessments and is used for high risk activities?

   a. Site notice board

   b. Hazard book

   c. Monitoring statement

   d. Method statement

6. Some substances are combustible. Which of the following are examples of combustible materials?

   a. Adhesives

   b. Paints

   c. Cleaning agents

   d. All of these

7. What is dermatitis?

   a. Inflammation of the skin

   b. Inflammation of the ear

   c. Inflammation of the eye

   d. Inflammation of the nose

8. Screens, netting and guards on a site are all examples of which of the following?

   a. PPE

   b. Signs

   c. Perimeter safety

   d. Electrical equipment

9. Which of the following are also known as scissor lifts or cherry pickers?

   a. Bench saws

   b. Hand-held power tools

   c. Cement additives

   d. Mobile elevated work platforms

10. In older properties the neutral electricity wire is which colour?

    a. Black

    b. Red

    c. Blue

    d. Brown

Chapter 2

Unit CSA–L1Core02

KNOWLEDGE OF TECHNICAL INFORMATION, QUANTITIES AND COMMUNICATION WITH OTHERS

LEARNING OUTCOMES

LO1: Know how to interpret construction related technical information

LO2: Know how to determine quantities of materials

LO3: Know how to relay information in the construction environment

LO4: Know how to communicate with others in the construction environment

INTRODUCTION

The aim of this chapter is to:

* show you the processes of passing on information

* show you the concepts of effective communication.

INTERPRETING CONSTRUCTION-RELATED TECHNICAL INFORMATION

Even quite simple construction projects will require documents. These provide you with the necessary information you will need to do the job. The documents are produced by a range of different people and each document has a different purpose. Together they give you the full picture of the job, from the basic outline through to the technical specifications.

Importance of documentation

In many industries a great deal of information is only ever stored electronically. This is not always an option in the construction industry. Many documents, such as working drawings, will need to be referred to on site. Detailed drawings of components that need to be made will have to be measured and checked before making joints, for example, in the workshop.

It is not always easy to store and look after working documents. The following advice is worth remembering:

* Always ensure you have the latest version of a document to work from before you begin to follow its instructions.

* If you are not going to need to use a document until later then get into the habit of storing it somewhere safe.

* Try to make sure that you do not leave documents lying around on site, where they could get lost or damaged.

* Try to make sure you always have a second copy of the document. You should keep this away from the site, in reserve, in case you lose your working copy.

* You should store any documents that you have used on a particular job at least until that job is completely finished.

* You might need to store the documents for some time after in case you need to refer back to them for repair and servicing.

Interpreting construction specifications

Obviously it would be impossible to put in all of the details in full, so symbols, hatchings and abbreviations are used to simplify the drawings. All of these symbols or hatchings are drawn to follow a British Standards-approved format, BS 1192. The symbols cover various types of brickwork and block work, as well as concrete, hard core and insulation, as can be seen in Fig 2.1.

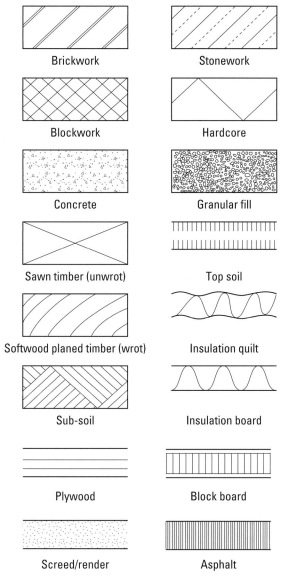

Figure 2.1 Symbols used on drawings

Common abbreviations

For the same reason, abbreviations are often used. Table 2.1 outlines some examples that you will need to become familiar with.

Abbreviation	Meaning
bwk	Brickwork
conc	Areas that will be concreted
dpc	Damp-proof course
fdn	Foundations
insul	Insulation
rwg	Rainwater gulleys
svp	Soil and vent pipe

Table 2.1

Types of documentation

Supporting information can be found in a variety of different types of documents. These include:

- drawings and plans
- programmes of work
- procedures
- specifications
- policies
- schedules
- manufacturers' technical information
- organisational documentation
- training and development records
- risk and method statements
- Construction (Design and Management) (CDM) Regulations
- Building Regulations.

Drawings and plans

Drawings are an important part of construction work. You will need to understand how drawings provide you with the information you need to carry out the work. The drawings show what the building will look like and how it will be constructed. This means that there are several different drawings of the building from different viewpoints.

Block plans

Block plans show the construction site and the surrounding area. Normally block plans are at a ratio of 1:2500 (usually in rural areas) and 1:1250 (usually in urban areas). This means that 1 mm on a block plan is equal to 2,500 mm or 1,250 mm on the ground.

Site plan

The site plan drawing shows what is basically planned for the site. It is an important drawing because it has been created in order to get approval for the project from planning committees or funding sources. In most cases the site plan is an architectural plan, showing the basic arrangement of buildings and any landscaping.

The site plan will usually show:

* directional orientation (i.e. the north point)

* location and size of the building or buildings

* existing structures

* clear measurements.

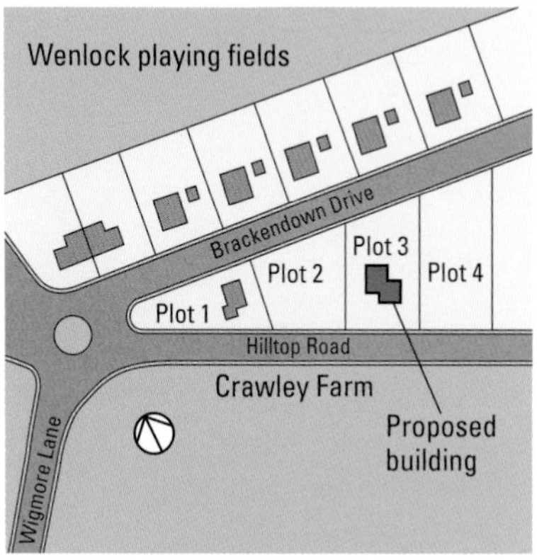

Figure 2.2 Block plan

General location

Location drawings show the site or building in relation to its surroundings. It will therefore show details such as boundaries, other buildings and roads. It will also contain other vital information, including:

* access   * sewers

* drainage   * the north point.

The scale will also be shown and the drawing will have a title. It will also be given a job or project number to help identify it easily, as well as an address, the date of the drawing and the name of the client. A version number will also be on the drawing with an amendment date if there have been any changes. You'll need to make sure you have the latest drawing.

Normally location drawings are either 1:500 or 1:200 (that is, 1 mm of the drawing represents 500 mm or 200 mm on the ground).

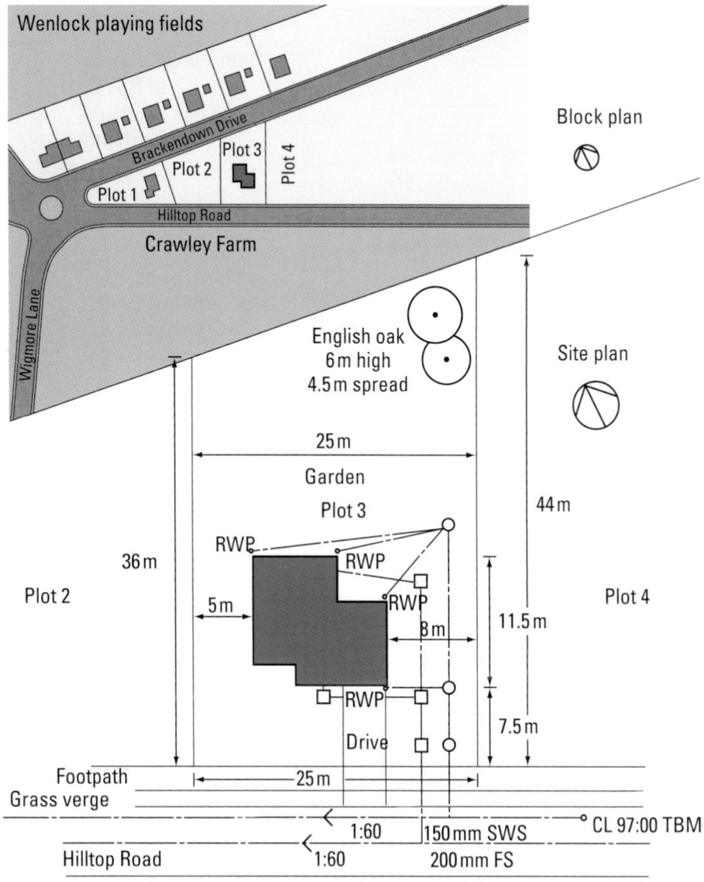

Assembly

These are detailed drawings that illustrate the different elements and components of the construction. They tend to be 1:20, 1:10 or 1:5 (1 cm of the drawing represents 20 mm, 10 mm or 5 mm on the ground). This larger scale allows more detail to be shown, to ensure accurate construction.

Figure 2.3 Location plan

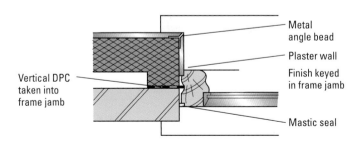

Figure 2.4 Assembly drawing

Sectional

These drawings aim to provide:

* vertical dimensions

* horizontal dimensions

* constructional details.

They can be used to show the height of ground levels, damp-proof courses, foundations and other aspects of the construction.

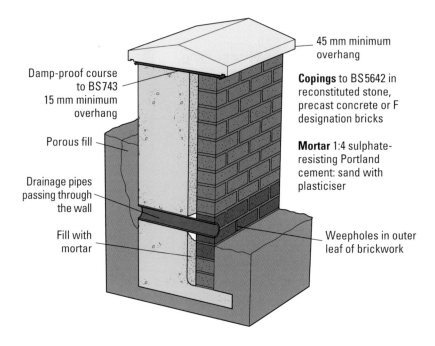

Damp-proof course to BS 743 15 mm minimum overhang

Porous fill

Drainage pipes passing through the wall

Fill with mortar

45 mm minimum overhang

Copings to BS 5642 in reconstituted stone, precast concrete or F designation bricks

Mortar 1:4 sulphate-resisting Portland cement: sand with plasticiser

Weepholes in outer leaf of brickwork

Figure 2.5 Section drawing of an earth retaining wall

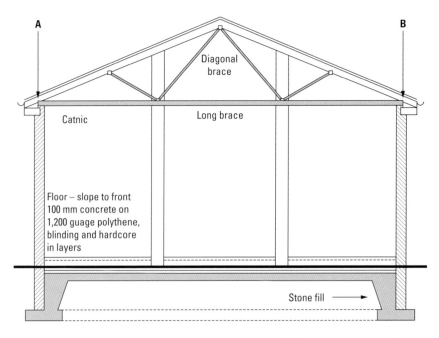

A

B

Diagonal brace

Catnic

Long brace

Floor – slope to front 100 mm concrete on 1,200 guage polythene, blinding and hardcore in layers

Stone fill

Figure 2.6 Section drawing of a garage

Detail drawings

These drawings show how a component needs to be manufactured. They are used to show the relationship between different components within the fabric of the building. For instance, an eaves detail would show rafters, wall plate, roof coverings, inner and outer masonry, insulation and much more. Details can be shown in various scales, but mainly 1:10, 1:5 and 1:1 (the same size as the actual component if it is small).

Orthographic projection (first angle)

First angle projection is a view that represents the side view, the front view and the plan view from above, as can be seen in Fig 2.8.

Serving hatch   Vertical section

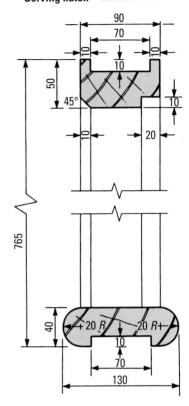

Figure 2.8 Detail drawing (measurements in mm)

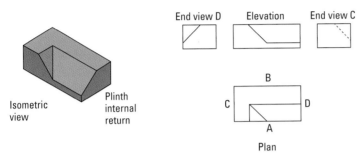

Figure 2.7 First angle projection

Isometric projection

Isometric projection is a way of representing three dimensional objects in two dimensions, as can also be seen in Fig 2.8. All horizontal lines are drawn at 30°.

Programmes of work

Programmes of work show the actual sequence of any work activities on a construction project. Part of the work programme plan is to show target times. They are usually shown in the form of a bar or Gantt chart (a special kind of bar chart), as can be seen in Fig 2.9.

DID YOU KNOW?

First angle is also known as European projection because Americans use third angle projection, which shows the views from a different position.

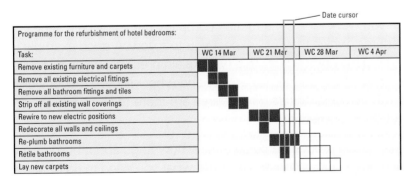

Figure 2.9 Single line contract plan Gantt chart

This figure shows the following:

* On the left hand side all of the tasks are listed – note this is ordered in the sequence of construction.

* On the right the blocks show the target start and end date for each of the individual tasks.

* The timescale can be either days, weeks or months.

Far more complex forms of work programmes can also be created. Fig 2.10 shows the construction of a house.

This more complex example shows the following:

* There are two lines – they show the target dates and actual dates. The actual dates are shaded, showing when the work actually began and how long it actually took.

* If this Gantt chart is kept up to date an accurate picture of progress and estimated completion time can be seen.

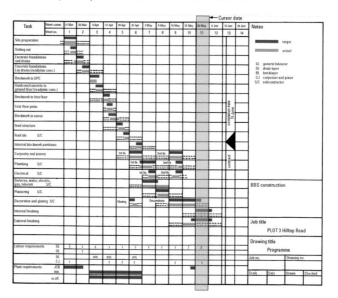

Figure 2.10 Gantt chart for the construction of a house

Procedures

When you work for a construction company there will be a series of procedures, which they will expect you to follow. A good example is the emergency procedure. This will explain precisely what is required in the case of an emergency on site and who will have responsibility to carry out particular duties. Procedures are there to show you the right way of doing something.

A construction procedure could outline how to go about building a wall or hanging a door, taking into account what you need to do beforehand, the materials and tools that are required, and the order in which you must carry out each step.

Another good example of a procedure is the procurement or buying procedure. This will outline:

* who is authorised to buy what, and how much individuals are allowed to spend

* any forms or documents that have to be completed when buying.

Specifications

In addition to drawings it is usually necessary to have documents known as specifications. These provide much more information, as can be seen in Fig 2.11.

The specifications give you a precise description. They will include:

● the address and description of the site

● on-site services (e.g. water and electricity)

● materials description, outlining the size, finish, quality and tolerances

● specific requirements, such as the individual who will authorise or approve work carried out

● any restrictions on site, such as working hours.

Policies

Policies are sets of principles or a programme of actions. The following are two good examples:

● The environmental policy outlines how the business goes about protecting the environment.

● The safety policy outlines how the business deals with health and safety matters and who is responsible for monitoring and maintaining it.

You will normally find both policies and procedures in site rules. These are usually explained to each new employee when they first join the company. Sometimes there may be additional site rules, depending on the job and the location of the work.

Schedules

Schedules are cross-referenced to drawings that have been prepared by an architect. They will show specific design information. Usually they are prepared for jobs that will be carried out regularly on site, such as:

● working on windows, doors, floors, walls or ceilings

● working on drainage, lintels or sanitary ware.

A schedule can be seen in Fig 2.12.

The schedule is very useful for:

● working out the quantities of materials needed

● ordering materials and components and then checking them against deliveries

● locating where specific materials will be used.

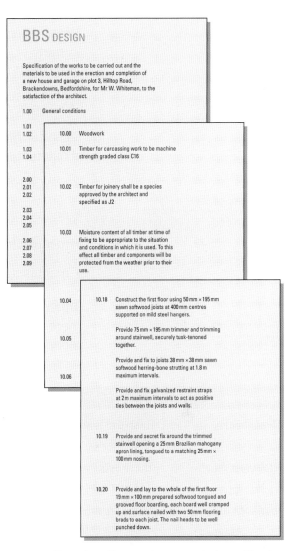

Figure 2.11 Extracts from a typical specification

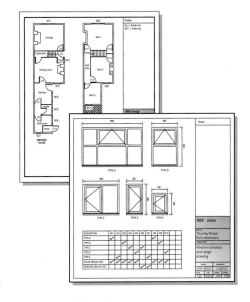

Figure 2.12 Typical windows schedule, range drawing and floor plans

Manufacturers' technical information

Almost everything that is bought to be used on site will come with a variety of information. The basic technical information provided will show what the equipment or material is intended to be used for, how it should be stored and any particular requirements it may have, such as handling or maintenance.

Technical information from the manufacturer can come from a variety of different sources:

* printed or downloadable data sheets

* printed or downloadable user instructions

* manufacturers' catalogues or brochures

* manufacturers' websites.

Organisational documentation

There is a huge potential list of organisational documentation and paperwork. Examples are outlined in Table 2.2. Visual examples can be seen in Fig 2.13 to 2.17.

Document	Purpose
Timesheet	Record of hours that you have worked and the jobs that you have carried out. This is used to help work out your wages and the total cost of the job.
Day worksheet	These detail work that has been carried out without providing an estimate beforehand. They usually include repairs or extra work and alterations.
Variation order	These are provided by the architect and given to the builder, showing any alterations, additions or omissions to the original job.
Confirmation notice	Provided by the architect to confirm any verbal instructions.
Daily report or site diary	These record things that might affect the project like detailed weather conditions, late deliveries or site visitors.
Orders and requisitions	These are order forms, requesting the delivery of materials.
Delivery notes	These are provided by the supplier of materials as a list of all materials being delivered. These need to be checked against materials actually delivered. The buyer will sign the delivery note when they are happy with the delivery.
Delivery record	These are lists of all materials that have been delivered on site.
Memorandum	These are used for internal communications and are usually brief.
Letters	These are used for external communications, usually to customers or suppliers.
Fax	Even though email is commonly used, the industry is still in favour of using faxes, as they provide an exact copy of an original document.

Table 2.2

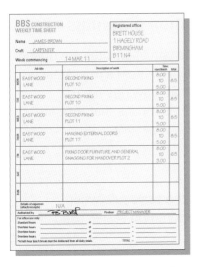

Figure 2.13 Timesheet

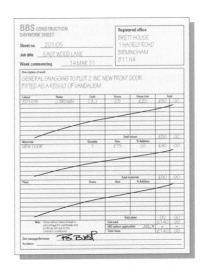

Figure 2.14 Day worksheet

Figure 2.15 Variation order

Figure 2.16 Confirmation notice

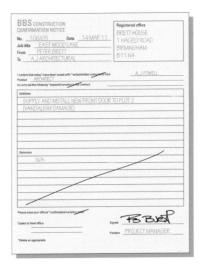

Figure 2.17 Daily report or site diary

Training and development records

Training and development is an important part of any job, as it ensures that employees have all the skills and knowledge that they need to do their work. Most medium to large employers will have training policies that set out how they intend to do this.

To make sure that they are on track and to keep records they will have a range of different documents. These will record all the training that an employee has undertaken.

Training can take place in a number of different ways:

* induction

* toolbox talks

* in-house training

* specialist training

* training or education leading to formal qualifications.

Scales used to produce construction drawings

When the plans for individual buildings or construction sites are drawn up they have to be scaled down so that they will fit on a manageable size of paper. It is important to remember that drawings are not sketches and that they are drawn to scale. This means that they are:

* exact and accurate

* in proportion to the real construction.

You can work out the dimensions by using the scale rule when measuring the drawings. There are several common scales used and the measurement is usually metric:

* 1:2500 – the drawing is 2,500 times smaller than the real object

* 1:100 – the drawing is 100 times smaller than the real object

* 1:50 – the drawing is 50 times smaller than the real object

* 1:20 – the drawing is 20 times smaller than the real object

* 1:10 – the drawing is 10 times smaller than the real object

* 1:5 – the drawing is 5 times smaller than the real object

* 1:2 – the drawing is 2 times smaller than the real object (also called 'half full size').

These drawings would clearly show the dimensions and these would be the actual measurements required (not the scaled-down measurements).

Selecting information

Various documents will provide you with the information you will need. The following examples show you how this works.

Location drawings and specifications

Location drawings are also known as block plans or site plans. The block plan shows the site presented as if you are looking down from above. It will show you where the site is in relation to other buildings and landmarks, such as roads.

The site plan will give you better detail of the site itself. It will contain measurements of the exact dimensions of the plot of land. It will also show you the routes of services and drainage.

The specifications are produced alongside the location drawings of the site. After giving you a brief description of the site, which includes the address, you will find:

* information about any services running into and through the site and whether or not they are connected or need to be connected

* whether there are restrictions about access or working hours on site

* materials that will be needed in order to carry out jobs on site (this will contain quite a lot of detail and will tell you what types of materials are needed, their size, quality and other technical details)

* information about the required workmanship, including what work needs to be done, what quality of work is expected and what the final finish should look like.

Schedules

Schedules can be quite big documents if you are working on a large site with lots of tasks to be done. There is usually a schedule for each different type of job. The schedules are there to record design information. They ensure that you do not accidentally use the wrong components or fittings.

Schedules cover all types of job, such as the types of doors, windows, joinery, heating components and other specific features.

Schedules usually have suitable drawings and usually a floor plan showing where the different features will appear.

DETERMINE QUANTITIES OF MATERIALS

Working out the quantity and cost of resources that are needed to do a particular job is, perhaps, one of the most difficult tasks. In most cases you or the company you work for will be asked to provide a price for the work. It is generally accepted that there are three ways of doing this:

* **An estimate** is an approximate cost produced from the information available before construction begins.

* **A quotation** is a fixed price.

* **A tender** is a bid for the job at your price.

These three ways of costing are very different and each of them has its own problems.

Checking deliveries of building materials

It is important that all deliveries are thoroughly checked as they arrive. Your suppliers will need to have access to the site. It is important to inform suppliers whether the site has any access problems. Construction materials usually arrive on site on large and heavy lorries, so always check if the ground they will have to cross is soft or uneven and warn them if this is the case. Trees and overhead wires could also be a problem, as could finding space to reverse and turn the lorry.

If you are expecting a delivery to arrive, you should be prepared for it. This means ensuring there is:

* clear access to the site

* somewhere ready to store the materials and equipment being delivered

* enough help on hand to assist moving the materials from the delivery point to the storage area.

There are two documents that you will need in order to check the delivery:

* your order, which is a purchase order or confirmation of the materials and equipment that you have ordered from the supplier

* a delivery note, which should be handed to you by the delivery driver.

The first thing to do is check that the two documents match. If they do then what is on the delivery truck should be what you ordered. You now need to check and tick off each item on the delivery note. It needs to be:

* the right specification   * the right quantity

* the right size   * undamaged.

You should not accept items that do not match these four points. You should not sign the delivery ticket or delivery note unless you are satisfied with what has been delivered.

Methods used to estimate quantities

Numerous factors determine the cost of a construction project, whatever its size, but getting the best price on the ideal quantities should not be guesswork. Some possible considerations, according to the size and significance of the project, would be:

* the availability of labour and materials

* the lead time on materials

* the economy and borrowing rates

* the time of year

* cost of plant

* the duration of the project – really large projects can go on for years.

Obviously experience means that you can more quickly estimate the quantities of materials that will be needed on particular (small) construction projects. This is also true of working out the best place to buy materials and how much the labour costs will be to get the job finished.

Many businesses will use the *Hutchins UK Building Blackbook* (published by Franklin-Andrews), which provides a construction cost guide. It breaks down all types of work and shows an average cost for each of them.

Computerised estimating packages are available, which will give a comprehensive detailed estimate that looks very professional. This will also help to estimate quantities and timescales.

The alternative is of course to carry out a numerical calculation. So it is important to have the right resources upon which to base these calculations. These could be working drawings, schedules or other documents.

Usually this involves making additions, subtractions, multiplications and divisions. In order to work out the amount of materials you will need for a construction project you will need to know some basic information:

* What does the job entail? How complex is it, and how much labour is required?

* What materials will be used?

* What are the costs of the materials?

Measurement

The standard unit for measurement is the metre (m). There are 100 centimetres (cm) and 1,000 millimetres (mm) in a metre. It is important to remember that drawings and plans have different scales, so these need to be converted to work out quantities of materials.

The most basic thing to work out is length, from which you can calculate perimeter, area and then volume, capacity, mass and weight, as can be seen in Table 2.3.

Measurement	Explanation
Length	This is the distance from one end to the other. This could be measured in metres or milimetres, depending on the job.
Perimeter	This is the total distance around the outside of a shape. For example, you might need to know the length of the perimeter around a site to work out how much security fencing you need before work starts. You can work out the perimeter by adding the lengths of each side of the shape together. For most jobs, perimeter will be measured in metres (see Fig 2.20).
Area	This is the amount of surface a shape covers. For example you might need to work out the area of a room or a wall to calculate what quantity of materials you will need. You can work out the area of a room by measuring the length and the width of the room and multiplying the two figures together. You can work out the area of a wall by measuring the length and the height of the wall and multiplying the two figures together. For most jobs, area will be measured in square metres (m^2) (see Fig 2.21).
Volume and capacity	This shows how much space is taken up by an object or room. You can work out the volume of a room by multiplying the width by the length and then by the height. For most jobs, volume will be measured in cubic metres (m^3).

Capacity works in exactly the same way as volume, but instead of showing the figure as cubic metres you may show it as litres (l). This is ideal if you are trying to work out the capacity of a water tank or a garden pond. |
| Mass or weight | Mass is measured usually in kilograms or in grams. Mass is the actual weight of a particular object, such as a brick. |

Table 2.3

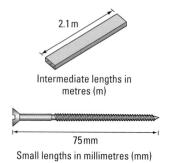

Intermediate lengths in
metres (m)

75mm

Small lengths in millimetres (mm)

Figure 2.18 Length in metres
and millimetres

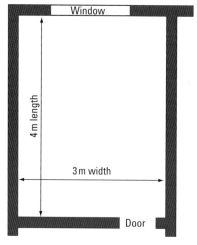

Figure 2.19 Measuring area and
perimeter

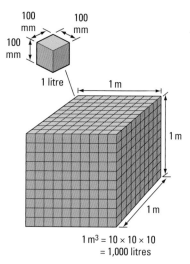

$1 m^3 = 10 \times 10 \times 10$
$= 1,000$ litres

Figure 2.20 Relationship between
volume and capacity

Formulae

These can appear to be complicated, but using formulae is essential for working out quantities of materials. Each of the formulae is related to different shapes. In construction you will often have to work out quantities of materials needed for odd shaped areas.

Area

To work out the area of a triangular shape, you use the following formula:

$$\text{Area (A)} = \text{Base (B)} \times \frac{\text{Height (H)}}{2}$$

So if a triangle has a base of 4.5 and a height of 3.5 the calculation is:

$$4.5 \times \frac{3.5}{2}$$

$$\text{Or } 4.5 \times 3.5 = \frac{15.75}{2} = 7.875\,m^2$$

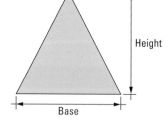

Figure 2.21 Triangle

Height

If you want to work out the height of a triangle you switch the formulae around. To give:

$$\text{Height} = 2 \times \frac{\text{Area}}{\text{Base}}$$

Perimeter

To work out the perimeter of a rectangle we use the formula:

$$\text{Perimeter} = 2 \times (\text{Length} + \text{Width})$$

It is important to remember this because you need to count the length and the width twice to ensure you have calculated the total distance around the object.

Circles

To work out the circumference or perimeter of a circle you use the formula:

$$\text{Circumference} = \pi\ (\text{pi}) \times \text{diameter}$$

π (pi) is always the same for all circles and is 3.142.

Diameter is the length of the widest part and is twice the radius.

If we know the circumference and need to work out the diameter of the circle the formula is:

$$\text{Diameter} = \frac{\text{circumference}}{\pi \ (\text{pi})}$$

For example if a circle has a circumference of 15.39 m then to work out the diameter:

$$\frac{15.39}{3.142} = 4.89\,\text{m}$$

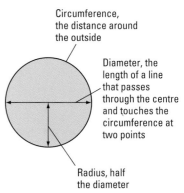

Circumference, the distance around the outside

Diameter, the length of a line that passes through the centre and touches the circumference at two points

Radius, half the diameter

Figure 2.22 Parts of a circle

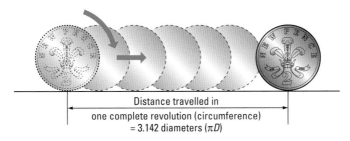

Distance travelled in one complete revolution (circumference) = 3.142 diameters (πD)

Figure 2.23 Relationship between circumference and diameter

Complex areas

Land, for example, is rarely square or rectangular. It is made up of odd shapes. Never be overwhelmed by complex areas, as all you need to do is break them down into regular shapes.

By accurately measuring the perimeter you can then break down the shape into a series of triangles or rectangles. All that is then necessary is to work out the area of each of the shapes within the overall shape and then add them together.

Shape	Area equals	Perimeter equals
Square	AA (or A multiplied by A)	4A (or A multiplied by 4)
Rectangle	LB (or L multiplied by B)	2(L+B) (or L plus B multiplied by 2)
Trapezium	$\dfrac{(A+B)H}{2}$ (or A plus B multiplied by H then divided by 2)	A+B+C+D

Shape	Area equals	Perimeter equals
Triangle	$\dfrac{BH}{2}$ (or B multiplied by H and then divided by 2)	A + B + C
Circle	πR^2 (or Pi (3.142) × R × R)	πD or 2πR (or Pi (3.142) × D or 2 × 3.142 × R)

Table 2.4 Calculating complex areas

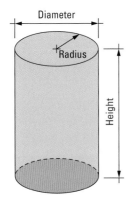

Figure 2.24 Cylinder

Volume

Sometimes it is necessary to work out the volume of an object, such as a cylinder or the amount of concrete needed. All that needs to be done is to work out the base area and then multiply that by the height.

For a concrete area, if a 1.2 m square needs 3 m of height then the calculation is:

$$1.2 \times 1.2 \times 3 = 4.32\,m^3$$

To work out the volume of a cylinder you need to know the base area × the height. The formula is:

$$\pi r^2 \times H$$

So if a cylinder has a radius (r) of 0.8 and a height of 3.5 m then the calculation is:

$$3.142 \times 0.8 \times 0.8 \times 3.5 = 7.038\,m^3$$

Pythagoras

Pythagoras' theorem is used to work out the length of the sides of right angled triangles. It states that:

In all right angled triangles the square of the longest side is equal to the sum of the squares of the other two sides (that is, the length of a side multiplied by itself).

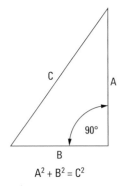

$$A^2 + B^2 = C^2$$

Figure 2.25 Pythagoras' theorem

Measuring materials

Using simple measurements and formulae can help you work out the amount of materials you will need. This is all summarised Table 2.5.

Material	Measurement
Timber	To work out the linear run of a cubic metre of timber of a given cross sectional area, divide a square metre by the cross sectional area of one piece.
Flooring	To work out the amount of flooring for a particular area metres² multiply the width of the floor by the length of the floor.
Stud walling, rafters and joists	Measure the distance that the stud partition will cover then divide that distance by a specified spacing and add 1. This will give you the number of spaces between each stud.
Fascias, barges and soffits	Measure the length and then add 10% for waste; however, this will depend on the nearest standard metric size of timber available.
Skirting, dado, picture rails and coving	You need to work out the perimeter of the room and then subtract any doorways or other openings. Again, add 10% for waste.
Bricks and mortar	Half brick walls use 60 bricks per metre squared and one brick walls use double that amount. You should add 5 per cent to take into account any cutting or damage. For mortar assume that you will need 1 kg for each brick.

Table 2.5 Working out materials required

How to cost materials

Once you have found out the quantity of materials necessary you need to find out the price of those materials. It is then simply a case of multiplying those prices by the amount of materials actually needed.

Materials and purchasing systems

Many builders and companies will have preferred suppliers of materials. Many of them will already have negotiated discounts based on their likely spending with that supplier over the course of a year. The supplier will be geared up to supply them at an agreed price.

In other cases builders may shop around to find the best price for the materials that match the specification. It is not always the case that the lowest price is necessarily the best. All materials need to be of a sufficient quality. The other key consideration is whether the materials are immediately available for delivery.

DID YOU KNOW?

Many businesses that fail do so as a result of not working out their costs properly. They may have plenty of work but they are making very little money.

CASE STUDY

LAING O'ROURKE

What you learned at school comes in handy for work

Joshua Richardson is an apprentice in his third year at Laing O'Rourke in Leeds.

'I got an A in Resistant Materials (Design and Technology) at school and then decided it was a good idea to apply for a joinery apprenticeship. I applied for a few apprenticeships and eventually got one with Laing O'Rourke. They gave me a phone interview, and then I had to go to an assessment day where we did things like spatial tests, group activities and a 10-minute presentation. I'm glad I got to do similar tests with my other applications, because if this was the first place I'd done them, maybe I wouldn't have made it.

You had to have GCSEs in Maths, English and Science at A–C. There's some people at college who didn't have these so they have to do Functional Skills on top of everything.

It's important because you do use your Maths and English skills at work every day.

Not only do you need to use your English for college work-based evidence, but also just talking to people. I found that having done talks and presentations at school, it really helped communication skills, and I can talk quite easily with different people now.

With Maths, you definitely need that every day. When you're doing any job, you need to work out how to use a measuring tape, and what kind of Maths you're going to have to use. You have to measure up properly and know each calculation you'll need. It's all on your tape, but you've got to think about it and add it all together, and it's especially important to get it right when you're cutting.

If you're not too good with numbers, you just have to practise it – it's something that can come to you, and not everyone gets it straight away. Every so often, a guy will shout out a couple of measurements and say, "Add that!" So it's not as if you get to pull out your calculator every time.'

It is vital that suppliers are reliable and that they have sufficient materials in stock. Delays in deliveries can cause major setbacks on site. It is not always possible to warn suppliers that materials will be needed, but a well-run site should be able to anticipate needed materials and put the orders in within good time.

Large quantities may be delivered direct from the manufacturer straight to site. This is preferable when dealing with items where consistency is essential.

RELAYING INFORMATION IN THE CONSTRUCTION ENVIRONMENT

Communication is all about passing on accurate information. No matter who you are communicating with, you must understand what is being asked. You also need to be able to give them a clear answer. Sometimes you do not have the information to hand or it is not your decision to make. In these cases you will need to take a message.

Whatever the situation, you always need to be positive and efficient. You also need to be clear. Poor communication and negative communication nearly always lead to confusion, delays and extra costs.

Basic content and requirements for recording a message

Message taking is an important skill. You are the way in which information is passed from one person to another. Providing you remember to pass the message on it may not need to be written down.

In many cases, however, messages can be complicated and these do need to be written down. In fact it is a wise plan to record the fact that the message was received in the first place:

* Note the date and time that the message was received.

* Clearly write down the actual message. Someone will have to read this, so make sure it is legible.

* Write down the person's name and their contact details.

If you are taking a message and you are in the site office then there may be a telephone answering pad to hand, or perhaps sticky notes.

If the questions are complicated it will be best to get them to call back or to promise them that the person with the knowledge will call them back.

PRACTICAL TIP

Many people who leave a message will tell you that it is urgent. It is not for you to decide whether it is or not. It is just down to you to pass the message on and mark it urgent if they have said so.

Positive and negative communication

Construction is an industry that relies on communication.

Positive communication means:

* being courteous and respectful when you are talking to others

* being considerate, particularly if the other person is under pressure

* listening to what others are saying

* being clear on key points

* keeping a sense of humour.

Showing these positive communication skills should mean that others will show positive communication towards you.

On the other hand, if you are:

* rude and disrespectful

* unwilling to listen or pay attention

* incapable of making a decision

* bad tempered

then you are communicating in a negative way. This could lead to confusion, arguments and problems.

Clear and effective communication

Communication in all types of work is essential. It needs to be clear and to the point, as well as accurate. Above all it needs to be a two-way process. This means that any communication you have with anyone must be understood. Think before communicating and never assume that someone understands you unless they have confirmed that they do. Negative communication or poor communication can damage the confidence that others have in you to do your job.

In construction work everything is about time and following strict instructions and specifications. Failing to communicate will always cause confusion, extra cost and delays. In an industry such as this these are unacceptable and very easy to avoid.

Good communication means efficiency and achievement.

COMMUNICATING WITH OTHERS IN THE CONSTRUCTION ENVIRONMENT

Communication can be split into two different types:

* **Verbal communication** includes face-to-face conversations, discussions in meetings or performance reviews and talking on the telephone.

* **Written communication** includes all forms of documents, from letters and emails to drawings and work schedules.

Each of these forms of communication needs to be clear, accurate and designed in such a way as to make sure that whoever has to use it or refer to it understands it.

Communicating in the appropriate way with others

Each construction job will require the services of a team of professionals. They will need to be able to work and communicate with one another. Each has different roles and responsibilities. They can be broken down into three particular groups:

* on site
* off site
* visitors.

These are described in Tables 2.6, 2.7 and 2.8.

On site

Role	Responsibilities
Apprentices	They can work for any of the main building services trades under supervision. They only carry out work that has been specifically assigned to them by a trainer, a skilled operative or a supervisor.
Skilled or trade operative	A specialist in a particular trade, such as bricklaying or carpentry. They will be qualified in that trade, or working towards their qualification
Unskilled operatives	Also known as labourers, these are entry level operatives without any formal training. They may be experienced on sites and will take instructions from the supervisor or site manager.
Building services engineers	They are involved in the design, installation and maintenance of heating, water, electrics, lighting, gas and communications. They work either for the main contractor or the architect and give instruction to building services operatives.
Building services operatives	They include all the main trades involved in installation, maintenance and servicing. They take instruction from the building services engineers and work with other individuals, such as the supervisor and charge-hand.
Sub-contractor	They carry out work on behalf of the main contractor and are usually specialist tradespeople or professionals, such as electricians. Essentially, they provide a service and are contracted to complete their part of the project.
Charge-hand	This person supervises a specific trade, such as carpenters and bricklayers.
Site manager	This person runs the construction site, makes plans to avoid problems and meet deadlines, and ensures all processes are carried out safely. They communicate directly with the client.
Supervisor	The supervisor works directly for the site manager on larger projects and carries out some of the site manager's duties on their behalf.
Health and safety officer	This person is responsible for managing the safety and welfare of the construction site. They will carry out inspections, provide training and correct hazards.

Table 2.6

Off site

Role	Responsibilities
Client	The client, such as a local authority, commissions the job. They define the scope of the work and agree on the timescale and schedule of payments.
Customer	For domestic dwellings, the customer may be the same as the client, but for larger projects a customer may be the end user of the building, such as a tenant renting local authority housing or a business renting an office. These individuals are most affected by any work on site. They should be considered and informed, with a view to them suffering as little disruption as possible.
Architect	They are involved in designing new buildings, extensions and alterations. They work closely with clients and customers to ensure the designs match their needs. They also work closely with other construction professionals, such as surveyors and engineers.

Consultant	Consultants such as civil engineers work with clients to plan, manage, design or supervise construction projects. There are many different types of consultant, all with particular specialisms.
Main contractor	This is the main business or organisation employed to head up the construction work. They organise the on-site building team and pull together all necessary expertise. They manage the whole project, taking full responsibility for its progress and costs.
Clerk of works	This person is employed by the architect on behalf of a client. They oversee the construction work and ensure that it represents the interests of the client and follows agreed specifications and designs.
Quantity surveyor	Quantity surveyors are concerned with building costs. They balance maintaining standards and quality against minimising the costs of any project. They need to make choices in line with Building Regulations. They may work either for the client or for the contractor, and clients and contractors may both have quantity surveyors on site.
Estimator	Estimators calculate detailed cost breakdowns of work based on specifications provided by the architect and main contractor. They work out the quantity and costs of all building materials, plant required and labour costs.
Supplier/wholesaler contracts manager	They work for materials suppliers or stockists, providing materials that match required specifications. They agree prices and delivery dates.

Table 2.7

Visitors

Site visitor	Role and responsibility
Training officers and assessors	These people work for approved training providers. They visit the site to observe and talk to apprentices and their mentors or supervisors. They assess apprentices' competence and help them to put together the paperwork needed to show evidence of their skills.
Building control inspector	This person works for the local authority to ensure that the construction work conforms to regulations, particularly the Building Regulations. They check plans, carry out inspections, issue completion certificates, work with architects and engineers and provide technical knowledge on site.
Water inspector	This person carries out checks of plumbing and drainage systems on construction sites.
Health and Safety Executive (HSE) inspector	An HSE inspector from the local authority can enter any workplace without giving notice. They will look at the workplace, the activities and the management of health and safety to ensure that the site complies with health and safety laws. They can take action if they find there is a risk to health and safety on site.
Electrical services inspector	Inspectors are approved by the National Inspection Council for Electrical Installation Contracting. They check all electrical installation has been carried out in accordance with legislation, particularly Part P of the Building Regulations.

Table 2.8

Maintaining good working relationships

It is important to have a good working relationship with colleagues at work. An important part of this is to communicate in a clear way with them. This helps everyone understand what is going on and what decisions have been made. It also means being clear. Most communication with colleagues will be verbal (spoken). Good communication means:

* cutting out mistakes and stoppages (saving money)

* avoiding delays

* making sure that the job is done right the first time and every time.

Equality and diversity in communication

Equality and diversity is not simply about treating everyone in the same way. It is actually recognising that people are different. Each of us is unique. This could mean that we might have a different culture, be of different ages or follow different religions. It might refer to our marital status or gender, our sexual orientation or our first language.

Figure 2.26 A water inspection

In all your actions and your communications you should:

* recognise and respect other people's backgrounds

* recognise that everyone has rights and responsibilities

* not harass or be offensive and use language or behaviour that discriminates.

You should also remember that not everyone's first language will be English so they may not understand everything or be able to communicate clearly with you. You might also find that some colleagues may have hearing impairments (or may not hear what you're saying because they are in a noisy environment). It's best to use simple language and check that both you and the person you're communicating with have understood what you need to know.

CASE STUDY

LAING O'ROURKE

Using writing and maths in the real world

Gary Kirsop, Head of Property Services says:

'People seem to think that trades are all about your hands, but it's more than that. You're measuring complicated things – all the trades need to have about the same technical level for planning, calculation and writing reports. You need that level to get through your exams for the future too. When you have one day a week in college, but four days a week working with customers in the real world, without communications skills, it would all fall apart. You have to understand that people come from different backgrounds and that they have their own communication modes. Having good GCSEs will really help you get by in the trade.'

TEST YOURSELF

1. If a drawing is at a scale of 1:500, each millimetre in the drawing represents how much on the ground?

    a. 1 m

    b. 1 mm

    c. 500 mm

    d. 500 m

2. What is the other term used to describe an orthographic projection?

    a. First angle

    b. Second angle

    c. Third angle

    d. Isometric

3. What is a variation order?

    a. A list of all materials that have been delivered to site

    b. A document showing work that has been carried out without a prior estimate

    c. A document that confirms any verbal instructions

    d. A document provided by the architect to the builder to show any changes to the original job

4. On a drawing, if you were to see the letters FDN, what would that mean?

    a. The signature of the architect

    b. Foundation Design Network

    c. Foundations

    d. Full distance

5. If a drawing is at a scale of 1:5, how many times smaller is the drawing than the real object?

    a. 5 times

    b. 50 times

    c. Half the size

    d. 500 times

6. Which of the following values is pi?

    a. 3.121

    b. 3.424

    c. 3.142

    d. 3.421

7. Which document is used to give detailed sets of requirements that cover the construction, features, materials and finishes?

    a. Work programme

    b. Purchase order

    c. Policy document

    d. Job specification

8. How do you work out the amount of flooring necessary for a room?

    a. Divide width by length

    b. Add width to length

    c. Multiply length by height

    d. Multiply width by length

9. Which individual on a typical site would sign off timesheets?

    a. Architect

    b. Site manager/supervisor

    c. Delivery person

    d. Customer

10. Which are the two main types of communication?

    a. Verbal and written

    b. Telephones and emails

    c. Meetings and memorandum

    d. Plans and faxes

Chapter 3

Unit CSA–L1Core03
KNOWLEDGE OF CONSTRUCTION TECHNOLOGY

LEARNING OUTCOMES

LO1: Know about foundation construction

LO2: Know about floor construction

LO3: Know about wall construction

LO4: Know about roof construction

LO5: Know about utilities and services within construction

LO6: Know about sustainability within construction

INTRODUCTION

The aim of this chapter is to:

* help you understand the range of building materials used within the construction industry

* help you understand their suitability to the construction of modern buildings.

* help you understand the role of sustainability in the construction industry

* help you to be aware of different construction methods

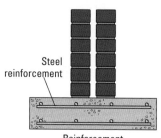

P X T

P

Distribution of load

T

45°

Width to be calculated

'P' greater than 'T' leads to shear failure

Steel reinforcement

Reinforcement used to reduce 'T'

Figure 3.1 Foundation properties

FOUNDATION CONSTRUCTION

Foundations are a primary element of a building, and form part of the substructure (the element of the building which is below ground and cannot be seen once the structure has been completed). Foundations spread the load of the superstructure (the visible part of the completed building), transferring it to the subsoil ground below. They provide structural stability and help to prevent damage to the building in the event of ground movement. They can range from a concrete strip to pre-cast reinforced concrete-driven piles.

Purpose of foundations

Foundations are designed to counteract factors such as ground movement, which could damage the building. It is important to work out the necessary width of foundation. This depends on the total load of the structure and the load-bearing capacity of the ground or subsoil on which the building is being constructed:

* Wide foundations are used when the construction is on weak ground, or the superstructure will be heavy.

* Narrow foundations are used when the subsoil is capable of carrying a heavy weight, or the building is a relatively light load.

The load that is placed on strip and pad foundations spreads into the ground at 45°. **Shear failure** will take place if the thickness of the foundations is less than the projection of the wall or column face on the edge of the foundations. This is what leads to subsidence (the ground under the structure sinking or collapsing).

As we will see in this section, the depth of the foundation is dependent on the load-bearing capacity of the subsoil. As a general rule of thumb, foundations should be 200 mm to 300 mm thick.

Different types of foundation

Generally there are several different types of foundation, which can be seen in Fig 3.2.

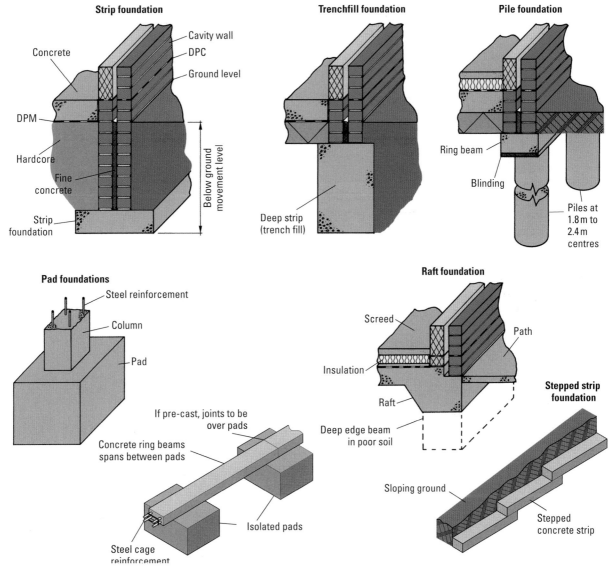

Figure 3.2 Foundation types

* The traditional **strip foundation** is quite narrow and tends to be used for low-rise buildings and dwellings. A thin strip of concrete is laid and then brick or block is built up to the DPC level. These can be reinforced where the ground is weak. They can also be stepped on sloping ground, in order to cut down on the amount of excavation needed. It can also be deep, which uses more concrete but reduces the amount of bricks used below ground level.

* Trench fill foundations are constructed by digging a narrow trench to the foundation depth and then filled with concrete. This reduces the labour and materials required to lay the foundation, as no bricks or blocks have to be laid into the trench. Trench fill:

* reduces the need to have a wide foundation

* reduces construction time

* speeds up the construction of the foundation.

* **Pads** tend to be used for structures that have either a concrete or a steel frame. The pads are placed to support the columns, which transfer the load of the building into the subsoil. Sometimes the pads are actually around the entire perimeter of the building to carry the load of the walls.

* **Pile foundations** tend to be used for high-rise buildings or where the subsoil is unstable. Holes are bored into the ground and filled with concrete or pre-cast concrete, steel or timber posts are driven into the ground. These piles are then spanned with concrete ring beams with steel reinforcement so that the load of the building is transferred deeper into the ground below. Pile foundations can be short or long depending on how high the building is or how bad the soil conditions are.

* **Raft foundations** are used when there is a danger that the subsoil is unstable. A large concrete slab reinforced with steel bars is used to outline the whole footprint of the building. It has an edge beam to take the load from the walls which is transferred over the whole raft. This means that the building effectively 'floats' on the ground surface on top of the concrete raft.

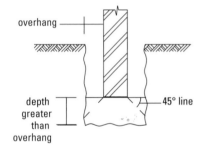

Figure 3.3 Unreinforced strip foundation

Selecting a foundation

One of the first things that a structural engineer will look at when they investigate a site is the nature of the soil, the ground conditions, the likelihood of ground movement and issues such as the water table (where groundwater begins).

Table 3.1 shows different types of subsoil and how they can affect the choice of foundation.

Subsoil type	Characteristics
Rock	High load bearing but there may be cracks or faults in the rock, which could collapse.
Granular	Medium to high load bearing and can be compacted sand or gravel. If there is a danger of flooding the sand can be washed away.
Cohesive	Low to medium load bearing, such as clay and silt. These are relatively stable, but there may be problems with water.
Organic	Low load bearing, such as peat and topsoil so organic material must be removed before starting the foundations. There is also a great deal of air and water present in the soil.

Table 3.1

The ground may move, particularly if the conditions are wet, extremely dry or there are extremes of temperature. Clay, for example, will shrink in the hot summer months and swell up again in the wet winter months. Frost can affect the water in the ground, causing it to expand (frost heave).

Ground movement is also affected by the proximity of trees and large shrubs. They will absorb water from the soil, which can dry out the subsoil. This causes the soil underneath the foundations to collapse. This may not be a problem until the tree or shrub is removed or new ones planted.

As we have seen above, the other key factor when selecting a foundation is the end use of the building:

* Strip foundations – this is the most common and cheapest type of foundation. It is a strip of concrete that runs under the load-bearing walls. The actual depth and width of the strip depends on the ground and the load from the building. They are used for low to medium rise domestic and industrial buildings, as the load bearing requirement will not be high.

* Piled foundations – tend to be used for high rise buildings or where subsoil is unstable if the ground directly under the building is weak or unstable then concrete or steel piles can be driven through this weak ground and into more solid ground beneath it. Reinforced concrete ring beams are placed over the piles as a direct support for the building.

* Raft foundations – these are very expensive and are only ever really used when the ground on which the building is being constructed is very soft. It is also sometimes used when the ground across the area is likely to react in different ways because of the weight of the building. In areas of the UK where there has been mining, for example, raft foundations are quite common, as the building could subside.

Concrete

Concrete is the most common material used for foundations as it is strong and durable. It is usually cast directly on site.

The concrete needs to be poured into the foundation with some care. The size of the foundation will usually determine whether the concrete is actually mixed on site or brought in, in a ready-mixed state, from a supplier. For smaller foundations a concrete mixer and wheelbarrows are usually sufficient. The concrete is then poured into the foundation using a chute (a long trough with a rounded bottom and open ends that directs the concrete to where it is needed).

Concrete consists of both fine and coarse aggregate, along with water.

Aggregates

Aggregates are basically fillers. Coarse aggregate is usually either crushed rock or gravel. The grains are 5mm or larger.

The fine aggregate fills up any gaps between the particles in the coarse aggregate.

Fine aggregate is usually sand that has grains smaller than 5mm.

Cement

Cement is an adhesive or binder. It is Portland stone, crushed, burnt and crushed again and mixed with limestone. The materials are powdered and then mixed together to create a fine powder, which is then fired in a kiln. It may be mixed with other materials for different purposes, such as creating masonry mortar.

Water

Potable water, which is water that is suitable for drinking, should be used when making concrete. The reason for this is that drinkable water has not been contaminated and it does not have organic material in it that could rot and cause the concrete to crack. The water mixes with the cement and then coats the aggregate. This effectively bonds everything together.

Additives

Additives, or admixtures, make it possible to control the setting time and other aspects of fresh concrete allowing you to have greater control over the concrete. They can:

* give you higher strength concrete
* provide protection against degradation of concrete or corrosion of reinforced concrete, which will weaken the structure
* speed up the time the concrete needs to set
* reduce the time the concrete takes to set
* provide protection against cracking as the concrete sets (by preventing shrinkage)
* improve the flow (workability) of the concrete
* improve the finish of the concrete
* provide hot or cold weather protection (a drop or rise in temperature can change the amount of time that concrete needs to set, so these admixtures compensate for that).

You might need this flexibility if, for example, the schedule or weather changes or the job has an unusual specification.

Figure 3.4 Reinforcement using steel bars (or mesh)

Reinforcement

Steel bars or mesh can be used to give the foundation additional strength and support. It can also help to stop the foundation from cracking. Concrete is a good material under direct weight loads, but where concrete foundations are wide and parts of them are under additional tension there is a danger they may crack.

Natural and artificial stone

Both of these products can be placed over the top of plain foundations hardcore fill that is put into the substructure to fill the gap up to the ground floor level. The synthetic stone weighs far less than natural stone. The other advantage with the synthetic products is that the foundations do not need to be as substantial.

FLOOR CONSTRUCTION

For most domestic buildings, floorboards or sheets are laid over timber joists. They then have a damp-proof membrane and over the top is solid concrete.

A floor is a level surface that provides some insulation and carries any loads on it (for example, from furniture) and then to transfer those loads.

Ground floors also have additional purposes. They need to stop moisture from entering the building from the ground. They also need to prevent plant or tree roots from entering the building.

Ground floors

For ground floors there are two options:

* Solid – in contact with the ground

* Suspended – does not touch the ground and spans between walls in the building (effectively there is a void beneath the floor)

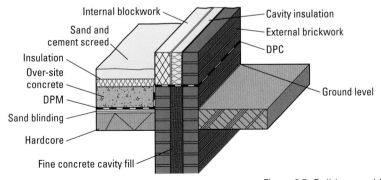

Figure 3.5 Solid ground floors

Floating floor

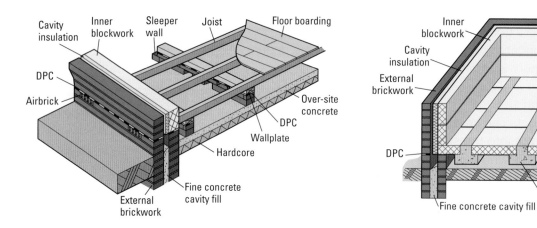

Figure 3.6 Suspended ground floors

The options for ground floors can be complicated because they need to perform several functions. While new builds don't tend to have timber joists and floorboards, extensions to existing buildings usually need to match existing construction styles. Suspended ground floors and traditional timber floors tend to be seen in older buildings. It is far more common to have solid ground floors, or to have timber floors over concrete floors, which are known as floating ground floors.

The key options are outlined in Table 3.2.

Type of floor	Construction and characteristics
Solid	The ground is compacted, and compacted hard core is used as the base, with a binding layer of sand, which is covered with a damp-proof membrane (dpm). A layer of insulation board, usually 100 mm thick, is then placed onto the dpm and concrete is poured on top. To provide a smooth finish for floor finishes a a cement and sand screed is applied, usually after the building has been made watertight.
Suspended	Timber – a similar process to a solid ground floor is carried out but then, on top of this, dwarf walls or sleepers are built. These are used to support the timber floor. Air bricks are also added to provide necessary ventilation. Joists are then spaced out along the dwarf walls. A damp-proof course is inserted under the floor joists and then floorboards or sheets placed on top of the joists. Beam and block – concrete beams and lightweight concrete slabs or blocks are used to create the basic flooring. The beams are evenly spaced across the foundation and gaps between the beams are filled with blocks to form the floor. The blocks and beams are then insulated and it is finished off with a cement screed.
Floating	This is a timber construction which goes over the top of a solid concrete floor. Bearers are put down and then the boarding or sheets are fixed to the bearers. The weight of the boards themselves hold them in place.

Table 3.2

Upper floors

Usually for dwellings timber is used for these suspended floors. In industrial buildings beam and block or concrete floor slabs tend to be used.

Timber suspended upper floor

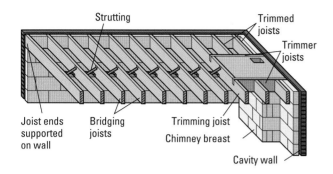

Concrete suspended upper floor

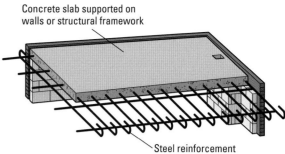

Figure 3.7 Upper floors

For dwellings, bridging joists (horizontal timbers that support the ceiling) are the most common joists used in suspended timber floors. These joists are supported at their ends by load-bearing walls. On the top of the joists, boarding or sheets provide the flooring for the room. Underneath the joists, plasterboard creates the basis of the ceiling for the room below.

When joists have to go into cavity walls (two walls with a hollow space between them) joist hangers are used (U-shaped metal brackets that are used to support the ends of floor joists). There are also complications when joists are in and around stairs and chimney breasts. Bridging joists are used so that these openings are not blocked. Openings in floors require the use of different types of joist called trimmers, trimming and trimmed joists. A trimmed joist is a shortened bridging joist. Any opening in a floor is treated in this way. When the span of a bridging joist exceeds 2 m then struts will be required in line with Building Regulations.

The voids between the floorboards and the plasterboard must be filled with insulation. This not only reduces heat loss, but can also reduce noise.

Concrete suspended floors are usually either cast on site or available as ready-cast units. They are effectively locked into the structure of the building by steel reinforcement. If the concrete floors are being cast on site then formwork is needed. Concrete floors are common because they offer greater load bearing capacity, have greater fire resistance and are more sound resistant.

KEY TERMS

Formwork

– this can also be known as shuttering. It is a temporary structure that supports and shapes wet concrete until it cures and is able to be self-supporting.

REED TIP

Try to remember that an effective team can produce more combined that a person can do on their own. Working with other people is also good for your personal well-being.

WALL CONSTRUCTION

Walls have a number of different purposes as they:

* hold up the roof

* provide protection against the elements

* keep the occupants of the building warm

* divide the building into rooms, providing privacy and different spaces.

External walls

Many buildings now have cavity walls which means:

* The outside wall is a wet one because it is exposed to the elements outside the building.

* The internal wall is dry but it needs to be kept separate from the outside wall by a cavity.

* The cavity or gap acts as a barrier against damp and also provides some heat insulation.

* The cavity can be completely filled or part-filled depending on the insulation value required by Building Regulations.

Internal walls

Internal walls divide up the space within the building. These do not have all of the demands of the external walls. They are less likely to be load bearing and they do not have to be insulated so are, therefore, thinner. (However, they are commonly insulated in areas such as the toilet or party walls in semi-detached or terrace construction.) They can be brick or block (particularly if they are load bearing), which is then covered with plaster. Alternatively they can be a timber or metal framework, known as stud work, which is covered plasterboard, to form a wall.

Different types of wall construction and structural considerations

In addition to walls being external or internal, they can also be classed as being load bearing or non-load bearing.

Internal walls can be either load bearing or non-load bearing. In both external and internal load bearing walls, any gaps or openings for windows or doors have to be bridged. This is achieved by using either arches or lintels. These support the weight of the wall above the opening.

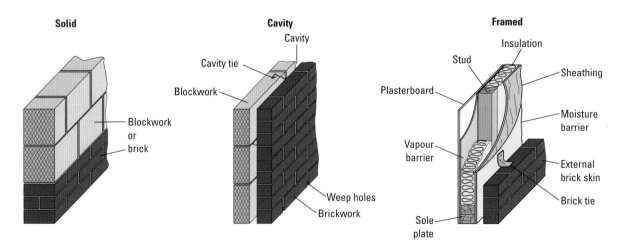

Figure 3.8 Some examples of external wall construction

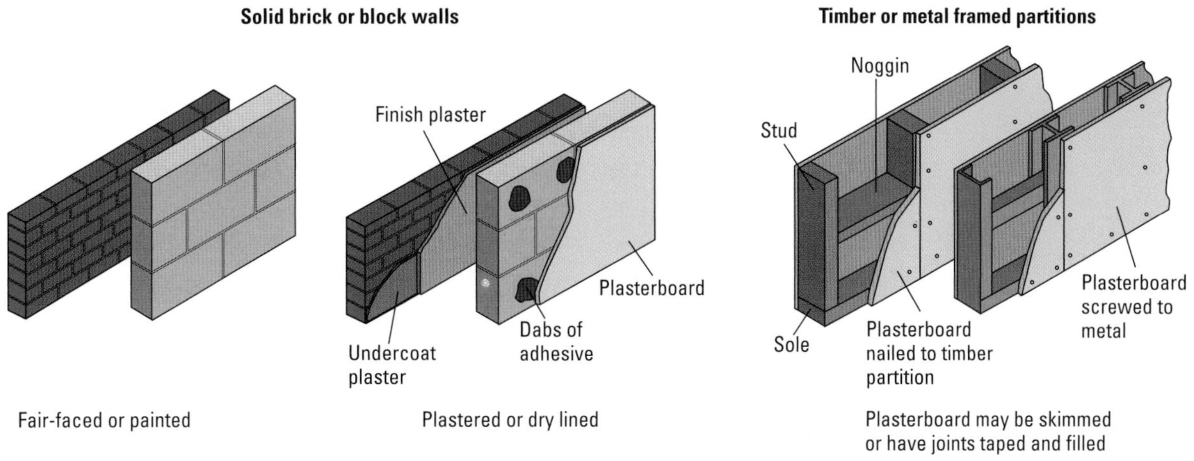

Solid brick or block walls

Timber or metal framed partitions

Finish plaster

Plasterboard

Dabs of adhesive

Undercoat plaster

Noggin

Stud

Sole

Plasterboard nailed to timber partition

Plasterboard screwed to metal

Fair-faced or painted

Plastered or dry lined

Plasterboard may be skimmed or have joints taped and filled

Figure 3.9 Some examples of internal wall construction

Solid walls

In modern builds construction of solid external walls is quite rare. External solid walls tend to be much thinner and made from lightweight blocks in modern builds. They will have some kind of waterproof surface over the top of them, which could be made of render, or plastic, metal or timber cladding.

Cavity walls

As we have seen, cavity walls have an outer and an inner wall and a cavity between them. Usually solid walling, or blockwork, is built up to ground level and then the cavity walling continues to the full height of the building. Alternatively a filled cavity wall is constructed up to ground level. Cavity walls are ideal for most buildings up to medium height.

Many industrial buildings have cavity walls for the lower part of the building and then have insulated steel panels for the top part of the building.

The usual technique is to have brick for the outer wall and an insulating block for the inner wall. The gap or cavity can then be partially filled with an insulation material.

Timber framed walls

Panels made of timber, or in some cases steel, are used to construct walls. They can either be load bearing or non-load bearing and can also be used for the outside of the building or for internal walling. Timber frames are also often clad in brickwork. The panels are solid structures and the spaces between the vertical struts (studs) and the horizontal struts (head or sole plates) are filled with insulation material.

Internal walls

Internal walls are either solid or framed. Solid walls can be made up from blocks or bricks. In many industrial buildings the blocks are actually exposed and can be left in their natural state or painted. In domestic buildings plasterboard is usually bonded to the surface and then plastered over to provide a smoother finish.

It is more common for domestic buildings to have timber or metal-framed internal walls, made from either timber or metal, which are known as stud partitions. These are exactly the same as other framed walling, but will usually have plasterboard fixed to them. They would then receive a skimmed coat of plaster to provide the smooth finish.

Damp-proof membrane (DPM) and damp-proof course (DPC)

Damp-proof membranes are installed under the concrete in ground floors in order to ensure that ground moisture does not enter the building. Effectively it waterproofs the building.

Damp-proof courses are a continuation of the damp-proof membrane. They are built into a horizontal course of either block or brickwork, which is a minimum of 150mm above the exterior ground level. DPCs are also designed to stop moisture from coming up from the ground, entering the wall and then getting into the building. The most common DPC is a polythene sheet damp-proof membrane, which comes in rolls the width of the blockwork or brickwork. In older buildings lead, bitumen or slate would have been used as a DPC.

ROOF CONSTRUCTION

In a country such as the UK, with a great deal of rain and snow, it makes sense for roofs to be pitched. Pitched means built at an angle. The idea is that the rain and snow falls down the angle and off the edge of the roof or into gutters rather than lying on the roof.

This is not to say that all roofs are pitched. In fact many domestic dwelling extensions have flat roofs. A great number of industrial buildings have entirely flat roofs. The problem with a flat roof is that it needs to be able to support itself, but as importantly it needs to be able to carry the additional weight of snow or rain. This means that large flat roofs may have to have steel sections (known as trusses) or even reinforced concrete and beams to increase their load-bearing capacity.

Roofs also provide stability to the walls by tying them together. As we will see, there are several different types of roof. These are usually identified by their pitch or shape.

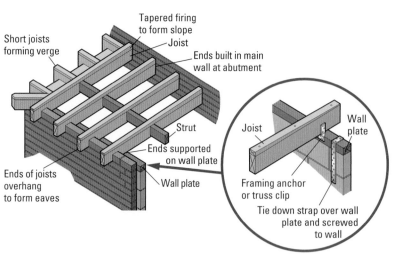

Short joists forming verge

Tapered firing to form slope

Joist

Ends built in main wall at abutment

Strut

Ends supported on wall plate

Ends of joists overhang to form eaves

Wall plate

Joist

Wall plate

Framing anchor or truss clip

Tie down strap over wall plate and screwed to wall

Figure 3.10 Flat roof structure

Types of roof construction

The roof is made up of the rafters and beams, Everything above the framework is regarded as a roof covering, such as slates, tiles and felt. Generally speaking, roofs are either pitched or flat. This will depend on the angle or slope of the roof.

Table 3.3 outlines some of the key characteristics of different types of roof.

Flat	This is a roof that has a slope of less than 10°. Generally flat roofs are used for smaller extensions to dwellings and on garages. Traditionally they would have had bitumen felt, although it is becoming more common for fibreglass to be used.	Figure 3.11
Mono-pitch	This is a roof that has a single sloping surface but is not fixed to another building or wall. The front and back walls could be different heights, or the other exposed surface of the roof is perpendicular.	Figure 3.12
Double pitched	This is a roof that has two differently angled slopes. Usually the upper part of the roof has a fairly shallow pitch or slope and the lower part of the roof has a steeper slope.	Figure 3.13
Couple roof	This is often called gable end and is one of the most common types of roof for dwellings. A gable is a wall with a triangular upper part. This supports the roof in construction using purlins. This means that the roof has two sloping surfaces, which come down from the ridge to the eaves.	Figure 3.14
Hipped roof	Hipped roofs have slopes on three or four four sides. There are also hipped roofs with single, straight gables.	Figure 3.15
Lean-to	A lean to is similar to a mono-pitched roof except it is abutted to a wall. The slope is greater than 10°. The higher part of the roof is fixed to a higher wall.	Figure 3.16

Table 3.3 Different types of roof

Roofing components

Each part of a roof has a specific name and purpose. Table 3.4 explains each of these individual features.

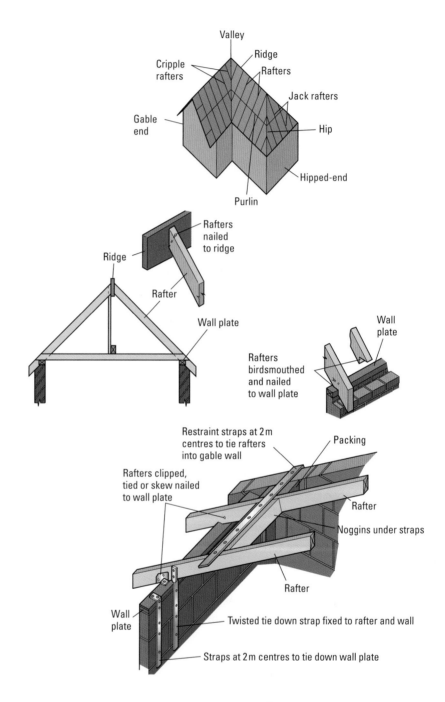

Figure 3.17 Traditional cut roof details

Roof feature	Description
Ridge	This is the top of the roof and the junction of the sloping sides. It is the apex, where the rafters meet.
Purlin	This is a beam that supports the mid-span section of rafters.
Firings	These are angled pieces of timber that are placed on the rafters to create a slope.
Batten	Roof battens are thin strips, usually of wood, which provide a fixing point for either roofing sheets or roof tiles.
Tile	These are artificial products that can be made from clay, concrete or plastic. They are placed in regular, overlapping rows and fixed to the battens.
Fascia	This is a horizontal board that closes and protects the rafter ends and provides a fixing for guttering. It is also decorative as it covers the rafter ends. It is fixed to the ends of the rafters at eaves level and is both a decorative feature and a fixing for rainwater goods..
Wall plate	This is a piece of horizontal timber that is placed at the top of a wall at eaves level. It provides a fixing for joists or rafters.
Bracings	Roof rafters may need to be braced to make them more rigid and stable. These bracings prevent the roof from buckling.
Felt	Roofing felt has two elements – it has a waterproofing agent (bitumen) and what is known as a carrier. The carrier can be either a polyester sheet or a glass fibre sheet. Roofing felt tends to be used for flat roofs and for roofs with a shallow pitch.
Slate	Slate roofing tiles are natural products and are usually fixed to timber battens with double nails. They have a lifespan of between 80 and 100 years.
Flashings	Wherever there is a joint or angle on a roof, a thin sheet of either lead or another waterproof material is added. In the past this tended always to be made from lead. Many different types of flashing can now be used but all have the role of preventing water penetrating into joints. Flashings are normally found where roofs abut a wall or where chimneys protrude through a roof.
Rafter	Roof rafters are the main structural components of the roof. They are the framework. They rest on supporting walls. The rafters are set at an angle on sloped roofs or horizontal on a flat roof.
Apex	The apex is the highest point of the roof, usually the ridge line.
Soffit	Soffits are the lower part, or overhanging part, of the eaves. In other words they are the underside of the eaves.
Bargeboard	This is a functional and ornamental feature, which is fixed to the gable end of a roof in order to hide the ends of roof timbers and to support the verge details.
Eaves	These are the area found at the foot of the rafter. They are not always visible as they can be flush. In modern construction, the eaves have two parts: the visible eaves projection and the hidden eaves projection.

Table 3.4

Roof coverings

There are many different types of materials that can be used to cover the roof. Even tiles and slates come in a wide variety of shapes and sizes, along with colours and different finishes.

In many cases the type of roof covering is determined by the traditional and local styles in the area. Local authorities usually want roof coverings that are not too far from the common style in the area. This does not stop manufacturers from coming up with new ideas, however, which can add benefits during construction and during the use of the building (such as better insulation properties).

Table 3.5 outlines some of the more common types of roof covering and describes their main characteristics and use.

Roof covering	Description
Felt  Figure 3.18	Felt is used as a waterproof barrier. Internal felt is rolled over the top of the rafters. The strips are overlapped to provide a permanent waterproof barrier. They are then battened down and another roof covering, such as slate or tile, placed over the top of them. For flat roofs, felt is used as the external roof covering and is covered in a waterproof material, such as bitumen.
Slate  Figure 3.19	Slate is a flat, natural substance, which is laid onto the battens with each slate tile overlapping the top of the slate in the row directly below it. The slate tiles are either nailed or hooked into place.
Tile Figure 3.20	There is a huge variety of roofing tiles, made from clay, ceramics or concrete. They are designed and moulded so that they overlap with one another and are fixed to the roof in a similar way to slate tiles.
Metals  Figure 3.21	There are many different types of metal roof covering, such as corrugated sheets, flat sheets, box profile sheets or even sheets that have a tile effect. The metal is galvanised and plastic coated to provide a durable and long-lasting waterproof surface.

Table 3.5

CASE STUDY

South
Tyneside Homes

South Tyneside Council's
Housing Company

How to impress in interviews

Andrea Dickson and Gillian Jenkins sit on the interview panels for apprenticeship applications at South Tyneside Homes.

'Interviews are all about the three Ps: Preparation, Presentation and Personality.

An applicant should turn up with some knowledge about the apprenticeship programme and the company itself. For example, knowing how long it is, that they have to go to college and to work – don't say, "I was hoping you'd tell me about it"! If they've done a bit of research, it will show through and work in their favour – especially if they can explain why it is that they want to work here.

It sets them up for the interview if they come in smartly dressed. We're not marking on that, but it does show respect for the situation. It's still a formal process and, although we try to make them feel at ease as much as we possibly can, there's no getting away from the fact that they're applying for a job and it is a formal setting.

The interviews are a chance to tell the company about themselves: what they do in their spare time, what their greatest achievements have been and why. Applicants should talk about what interests them; for example, are they really interested in becoming a joiner or is that something their parents want them to do? An apprenticeship has to be something they want to do – if they have enthusiasm for the programme, then they'll fly through it. If not, it's a very long three to four years. Without that passion for it, the whole process will be a struggle; they'll come in late to work and even fail exams.

We also talk to them about any customer service experiences they've had, working in a team, project working (for example, a time you had to complete a task and what steps you took), as well as asking some questions about health and safety awareness.'

SUPPLY OF UTILITIES AND SERVICES

Most but not all dwellings and other structures are connected in some way to a wide range of utilities and services. In the majority of cities, towns and villages structures are connected to key utilities and services, such as a sewer system, potable (drinking) water, gas and electricity. This is not always the case for more remote structures, however.

Whenever construction work is carried out, whether it is on an existing structure or a new build, the supply of utilities and services or the linking up of these parts of the **infrastructure** are very important. Often they will require the services of specialist engineers from the **service provider.**

KEY TERMS

Infrastructure

– these are basic facilities, such as a power supply, a road network and a communication link.

Service provider

– these are companies or organisations that provide utilities, such as gas, water, communications or electricity.

Table 3.6 outlines the main utilities and services that are provided to most structures.

Utility or service	Description
Drainage	Drainage is delivered by a range of water and sewerage companies in the UK. They are responsible for ensuring that surface water can drain away into their system.
Waste water and sewerage	Any waste water and sewage generated by the occupants of a structure needs to have the necessary pipework to link it to the main sewerage system. It is then sent to a sewage treatment works via the pipework. If there is no connection to mains sewerage, the building may have a septic tank, which is a small-scale, self-contained sewage treatment system.
Water	Each structure should be linked to the water supply that provides wholesome, potable drinking water. The pipework linking the structure to the water supply needs to be protected to ensure that backflow from any other source does not contaminate the system.
Gas	Each area has a range of different gas suppliers. This is delivered via a service pipe from the main system into the structure. Areas that do not have access to the main gas supply system use gas contained in cylinders.
Electricity	The National Grid provides electricity to a variety of different electricity suppliers. It is the National Grid that operates and maintains the cabling. There are around 28 million individual electricity customers in the UK.
Communications (telephone, data, cable)	There are several ways in which telecommunications can be linked to a structure. Traditional telephone poles hold up copper cables and not only provide telephone but also internet access to structures. In cities and many of the larger towns poles are being replaced by cables that are fibre optic and run underground. These are then linked to each individual structure.
Ducting (heating and ventilation)	Heating and ventilation engineers install and maintain duct work. The complex systems are known as HVAC. These systems can transfer air for heating or cooling of the structure. The overall system can also provide hot and cold water systems, along with ventilation.

Table 3.6 Services and utilities

KEY TERMS

HVAC

– this is an abbreviation for 'heating, ventilation and air-conditioning'. This has been a service provided to many industrial buildings for a number of years, but it is now becoming more common in domestic dwellings, particularly new developments.

SUSTAINABILITY AND INCORPORATING SUSTAINABILITY INTO CONSTRUCTION PROJECTS

Carbon is present in all fossil fuels, such as coal or natural gas. Burning fossil fuels releases carbon dioxide, which is a greenhouse gas linked to climate change.

Energy conservation aims to reduce the amount of carbon dioxide in the atmosphere. The idea is to do this by making buildings better insulated and, at the same time, make heating appliances more efficient. It also means attempting to generate energy using renewable and/or low or zero carbon methods.

According to the government's Environment Agency, sustainable construction is all about using resources in the most efficient way. It also means cutting down on waste on site and reducing the amount of materials that have to be disposed of and put into landfill.

In order to achieve sustainable construction the Environment Agency recommends:

* reducing construction, demolition and excavation waste that needs to go to landfill

* cutting back on carbon emissions from construction transport and machinery

* responsibly sourcing materials

* cutting back on the amount of water that is wasted

* making sure construction does not have an impact on biodiversity.

Sustainable construction and incorporating it into construction projects

Recently the idea of sustainable construction has focused on ensuring that the building is not only of good quality and affordable, but also that it is efficient.

Sustainable construction also means having the least negative environmental impact. So this means minimising the use of raw materials, energy, land and water. This is not only during the construction phase but also for the lifetime of the building.

Finite and renewable resources

We all know that resources such as coal and oil will eventually run out. These are examples of finite resources.

Oil is not just used as fuel – it is in plastic, dyes, lubricants and textiles. All of these are used in the construction process.

Renewable resources are those that are produced either by moving water, the sun or the wind. They include materials that come from plants, such as biodiesel, or the oils used to make adhesives.

The construction process itself is only part of the problem. It is also the longer term impact and demands that the building will have on the environment. This is why there has been a drive towards sustainable homes and there is a Code for Sustainable Homes (an environmental assessment method for rating and certifying the performance of new homes).

Figure 3.22 Most modern new-builds follow sustainable principles

Construction and the environment

In 2010 construction, demolition and excavation produced 20 million tonnes of waste that had to go into landfill. The construction industry is also responsible for most illegal fly tipping (illegally dumping waste). In any year the Environment Agency responds to around 350 serious pollution incidents caused as a result of construction.

Regardless of the size of the construction job, everyone working on the project is responsible for the impact they have on the environment. Good site layout, planning and management can help reduce these problems.

Sustainable construction helps to encourage this because it means managing resources in a more efficient way, reducing waste and reducing your **carbon footprint.**

Architecture and design

The Code for Sustainable Homes Rating Scheme was introduced in 2007. Many local authorities have instructed their planning departments to encourage sustainable development. This begins with the work of the architect who designs the building.

Local authorities ask that architects and building designers:

* ensure the land is safe for development – that if it is contaminated this is dealt with first

* ensure there is access to and protection of the natural environment – this helps ensure biodiversity and tries to create open spaces for local people

* reduce the negative impact on the local environment – any buildings keep noise, air, light and water pollution down to a minimum

* conserve natural resources and cut back carbon emissions – this includes use of energy, materials and water during construction and the life of the building

* Ensure comfort and security – good access, close to public transport, safe parking and protection against flooding.

Figure 3.23 Sustainable developments aim to be pleasant places to live

Using locally managed resources

The construction industry imports nearly 6 million cubic metres of sawn wood each year. However there is plenty of scope to use the many millions of cubic metres of timber produced in managed forests in the UK, particularly in Scotland.

Local timber can be used for a wide variety of different construction projects:

* softwood – including pines, firs, larch and spruce – for panels, decking, fencing and internal flooring

* hardwood – including oak, chestnut, ash, beech and sycamore – for a wide variety of internal joinery.

Using local materials reduces transportation costs and time, minimises the project's carbon footprint and means that there is less chance for the materials to be damaged in transit.

Eco-friendly, sustainable manufactured products and environmentally resourced timber

There are now many suppliers that offer sustainable building materials as a green alternative. Some tiles, for example, are now made from recycled plastic bottles and stone particles.

Figure 3.24 Window frames made from timber

There is now a National Green Specification database of all environmentally friendly building materials. This provides a checklist where it is possible to compare specifications of sustainable products to traditionally manufactured products, such as bricks.

Simple changes can be made, such as using timber or ethylene-based plastics instead of UPVC window frames to ensure a building uses more sustainable materials.

As we have seen, finding locally managed resources, such as timber, makes sense in terms of cost and in terms of protecting the environment. There are always alternatives to the use of traditional resources that could affect the environment.

The Timber Trade Federation produces a timber certification system. This ensures that wood products are labelled to show that they are produced in sustainable forests.

Around 80 per cent of all the softwood used in construction comes from Scandinavia or Russia. Another 15 per cent comes from the rest of Europe, or even North America. The remaining 5 per cent comes from tropical countries, and is usually sourced from sustainable forests.

Alternative methods of building

The most common type of construction is, of course, brick and blockwork. However there are plenty of other options:

* timber frame – using green oak

* insulated concrete formwork – where a polystyrene mould is filled with reinforced concrete

DID YOU KNOW?

www.recycledproducts. org.uk has a long list of recycled surfacing products, such as tiles, recycled wood and paving and details of local suppliers.

Figure 3.25 Timber Certification System

85

* structural insulated panels – where buildings are made up of rigid building boards rather like huge sandwiches

* modular construction – this uses similar materials and techniques to standard construction, but the units are built off site and transported ready-constructed to building site where they are connected together.

Figure 3.26 Green roofing

Figure 3.27 Flooring made from cork

KEY TERMS

Biodegradable

– this material will more easily break down when it is no longer needed. This breaking down process is done by micro-organisms.

Organic

– these are natural substances, usually extracted from plants.

There are alternatives to traditional flooring and roofing, all of which are greener and more sustainable. Green roofing has become an increasing trend in recent years. Metal roofs made of steel, aluminium and copper use a high percentage of recycled material. Solar roof shingles, or solar roof laminates, while expensive, decrease the cost of electricity and related heating costs of the dwelling. Some buildings even have a waterproof membrane, which is covered with a growing medium and planted with vegetation like sedum plants. This provides additional insulation, absorbs air pollution, helps to collect and process rainwater and keeps the roof surface temperature down.

Just as roofs are becoming greener, so too are the options for flooring. The use of bamboo, eucalyptus and cork is becoming more common. A new version of linoleum has been developed with **biodegradable**, **organic** ingredients. Some buildings are also using sustainable alternatives to traditional timber floorboards and joists, and these can be coloured, stained or patterned.

An increasing trend has been for what is known as off-site manufacture (OSM). European OSM businesses, particularly those in Germany, have built over 100,000 houses. The entire house is manufactured in a factory and then assembled on site. Walls, floors, roofs, windows and doors with built-in electrics and plumbing, all arrive on a lorry. Some manufacturers even offer completely finished dwellings, including carpets and curtains. Many of these modular buildings are designed to be far more energy efficient than traditional brick and block constructions. Many come ready fitted with heat pumps, solar panels and triple-glazed windows.

Energy efficiency and incorporating it into construction projects

Energy efficiency is all about using less energy to provide the same level of output. Governments are working towards the world's energy needs by 30 per cent before 2050. This means producing more energy efficient buildings. It also means using energy efficient methods to produce materials and resources needed to construct buildings.

Building Regulations

In terms of energy conservation, the most important UK law is the Building Regulations 2010, particularly Part L. The Building Regulations:

* list the minimum efficiency requirements

* provide guidance on compliance, the main testing methods, installation and control

* cover both new dwellings and existing dwellings.

A key part of the regulations is the Standard Assessment Procedure (SAP), which measures or estimates the energy efficiency performance of buildings.

Local planning authorities also now require that all new developments generate at least 10 per cent of their energy from renewable sources. This means that each new project has to be assessed one at a time.

Energy conservation

By law, each local authority is required to reduce carbon dioxide emissions and to encourage the conservation of energy. This means that everyone has a responsibility in some way to conserve energy:

* Clients, along with building designers, are required to include energy efficient technology in the build.

* Contractors and sub-contractors have to follow these design guidelines. They also need to play a role in conserving energy and resources when actually working on site.

* Suppliers of products are required by law to provide information on energy in the production of their products.

In addition, new energy efficiency schemes and Building Regulations cover the energy performance of buildings. Each new build is required to have an Energy Performance Certificate. This rates a building's energy efficiency from A (which is very efficient) to G (which is very inefficient).

Some building designers have also begun to adopt other voluntary ways of attempting to protect the environment. These include BREEAM, which is an environmental assessment method, and the Code for Sustainable Homes, which is a certification of sustainability for new builds.

Figure 3.28 The Energy Saving Trust encourages builders to use less wasteful building techniques and more energy efficient construction

High, low and zero carbon

When we look at energy sources, we consider their environmental impact in terms of how much carbon dioxide they release. Accordingly, energy sources can be split into three different groups:

* high carbon – those that release a lot of carbon dioxide

* low carbon – those that release some carbon dioxide

* zero carbon – those that do not release any carbon dioxide.

Some examples of high carbon, low carbon and zero carbon energy sources are given in Table 3.7

High carbon energy source	Description
Natural gas or LPG	Piped natural gas or liquid petroleum gas stored in bottles
Fuel oils	Domestic fuel oil, such as diesel
Solid fuels	Coal, coke and peat
Electricity	Generated from non-renewable sources, such as coal-fired power stations
Low carbon energy source	
Solar thermal	Panels used to capture energy from the sun to heat water
Solid fuel	Biomass such as logs, wood chips and pellets
Hydrogen fuel cells	Convert chemical energy into electrical energy
Heat pumps	Convert low temperature heat into higher temperature heat
Combined heat and power (CHP)	Generates electricity as well as heat for water and space heating
Combined cooling, heat and power (CCHP)	A variation on CHP that also provides a basic air conditioning system
Zero carbon energy	
Electricity/wind	Uses natural wind resources to generate electrical energy
Electricity/tidal	Uses wave power to generate electrical energy
Hydroelectric	Uses the natural flow of rivers and streams to generate electrical energy
Solar photovoltaic	Uses solar cells to convert light energy from the sun into electricity

Table 3.7 High, low and zero carbon energy sources

It is important to try to conserve non-renewable energy so that there will be sufficient fuel for the future. The idea is that finite sources of fuel should last as long as is necessary to completely replace it with renewable sources, such as wind or solar energy.

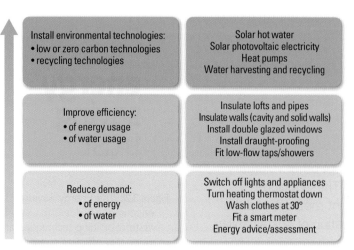

Figure 3.29 Working towards reducing carbon emissions

Alternative heating sources

There are several new ways in which we can harness the power of water, the sun and the wind to provide us with new heating sources. All of these systems are considered to be far more energy efficient than traditional heating systems, which rely on gas, oil, electricity or other fossil fuels.

Solar thermal

At the heart of this system is the solar collector, which is often referred to as a solar panel. The idea is that the collector absorbs the sun's energy, which is then converted into heat. This heat is then applied to the system's heat transfer fluid.

The system uses a differential temperature controller (DTC) that controls the system's circulating pump when solar energy is available and there is a demand for water to be heated.

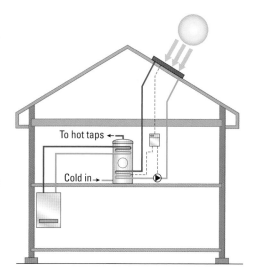

Figure 3.30 Solar thermal hot water system

In the UK, due to the lack of guaranteed solar energy, solar thermal hot water systems often have an auxiliary heat source, such as an immersion heater.

Biomass (solid fuel)

Biomass stoves burn either pellets or logs. Some have integrated hoppers that transfer pellets to the burner. Biomass boilers are available for pellets, woodchips or logs. Most of them have automated systems to clean the heat exchanger surfaces. They can provide heat for domestic hot water and space heating.

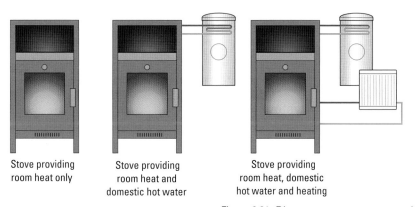

Stove providing room heat only

Stove providing room heat and domestic hot water

Stove providing room heat, domestic hot water and heating

Figure 3.31 Biomass stoves output options

Heat pumps

Heat pumps convert low temperature heat from air, ground or water sources to higher temperature heat. They can be used in ducted air or piped water **heat sink** systems.

There are a variety of different arrangements for each of the three main systems:

* Air source pumps operate at temperatures down to minus 20°C. They have units that receive incoming air through an inlet duct.

* Ground source pumps operate on **geothermal** ground heat. They use a sealed circuit collector loop, which is buried either vertically or horizontally underground.

* Water source systems can be used where there is a suitable water source, such as a pond or lake.

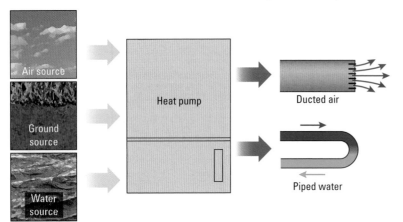

Figure 3.32 Heat pump input and output options

The heat pump system's efficiency relies on the temperature difference between the heat source and the heat sink. Special tank hot water cylinders are part of the system, giving a large surface-to-surface contact between the heating circuit water and the stored domestic hot water.

Combined heat and power (CHP) and combined cooling heat and power (CCHP) units

These are similar to heating system boilers, but they generate electricity as well as heat for hot water or space heating (or cooling). The heart of the system is an engine or gas turbine. The gas burner provides heat to the engine when there is a demand for heat. Electricity is generated along with sufficient energy to heat water and to provide space heating.

CCHP systems also incorporate the facility to cool spaces when necessary.

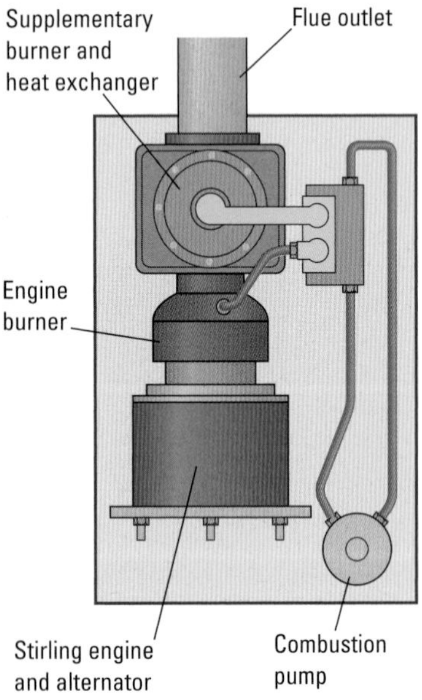

Wind turbines

Freestanding or building-mounted wind turbines capture the energy from wind to generate electrical energy. The wind passes across rotor blades of a turbine, which causes the hub to turn. The hub is connected by a shaft to a gearbox. This increases the speed of rotation. A high speed shaft is then connected to a generator that produces the electricity.

Solar photovoltaic systems

A solar photovoltaic system uses solar cells to convert light energy from the sun into electricity. The solar cells are usually made of silicon and are semi-conductors. The sunlight hits the solar cells and photons are absorbed. This causes negatively charged electrons in the cell to detach from their atoms and flow through the cell to create electricity. The electricity is direct current (dc). The dc current is then converted by an inverter to alternating current (ac), which is the type of current used for mains electricity.

Figure 3.33 Example of a MCHP (micro combined heat and power) unit

Energy ratings

Energy rating tables are used to measure the overall efficiency of a dwelling, with rating A being the most energy efficient and rating G the least energy efficient.

Alongside this is an environmental impact rating (see Fig 3.38). This measures the dwelling's impact on the environment in terms of how much carbon dioxide it produces. Again, rating A is the highest, showing it has the least impact on the environment, and rating G is the lowest.

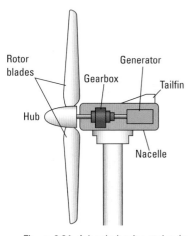

Figure 3.34 A basic horizontal axis wind turbine

A standard assessment procedure (SAP) is used to place the dwelling on the energy rating table. This will take into account:

* the date of construction, the type of construction and the location

* the heating system

* insulation (including cavity wall)

* double glazing.

The ratings are used by local authorities and other groups to assess the energy efficiency of new and old housing, and must be provided to potential purchasers when houses are sold.

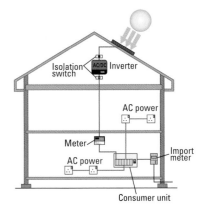

Figure 3.35 A basic solar photovoltaic system

Preventing heat loss

Most old buildings are under-insulated and would benefit from additional insulation, whether this is ceilings, walls or floors.

The measurement of heat loss in a building is known as the U Value. It measures how well parts of the building transfer heat. Low U Values represent high levels of insulation. U Values are becoming more important as they form the basis of energy and carbon reduction standards.

By 2016 all new housing is expected to be Net Zero Carbon. This means that the building should not be contributing to climate change.

Many of the guidelines are now part of Building Regulations (Part L). They cover:

* insulation requirements

* openings, such as doors and windows

* solar heating and other heating

* ventilation and air conditioning

* space heating controls

* lighting efficiency

* air tightness.

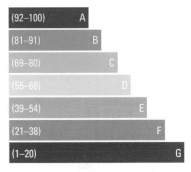

Figure 3.36 SAP energy efficiency rating table. The ranges in brackets show the percentage energy efficiency for each banding

Building design

UK households spend £2.4bn every year just on lighting. One of the ways of tackling this cost is to use energy saving lights, but also to maximise natural lighting. For the construction industry this means:

* increased window size

* orientating building angles to make the most of sunlight – south facing windows maximise sunlight in winter and limit overheating in the summer

* considering window design by using windows with a variety of different types of opening to allow ventilation.

Solar tubes are another way of increasing light. These are small domes on the roof, which collect sunlight and then direct it through a tube (which is reflective). It is then directed through a diffuser in the ceiling to spread light into the room.

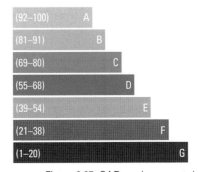

Figure 3.37 SAP environmental impact rating table

TEST YOURSELF

1. In which of the following types of buildings is a traditional strip foundation used?

   a. High rise

   b. Medium rise

   c. Low rise

   d. Industrial buildings

2. Which of the following is a reason for using a raft foundation?

   a. The subsoil is rock

   b. The subsoil is unstable

   c. The subsoil is stable

   d. The access to the site allows it

3. What holds down a floating floor?

   a. Nails and screws

   b. Adhesives

   c. Blocks

   d. Its own weight

4. What is another term for formwork?

   a. Shuttering

   b. Cavity

   c. Joist

   d. Boarding

5. What is the minimum distance the DPC should be above ground level?

   a. 50 mm

   b. 100 mm

   c. 150 mm

   d. 200 mm

6. A roof is said to be flat if it has a slope of less than how many degrees?

   a. 5

   b. 10

   c. 15

   d. 20

7. What shape is the upper part of a gable end?

   a. Rectangular

   b. Semi-circular

   c. Square

   d. Triangular

8. What do you call the horizontal timber that is placed at the top of a wall at eaves level in a roof, to hold the ends of joists or rafters?

   a. Fascia

   b. Bracings

   c. Wall plate

   d. Batten

9. What happens to the majority of construction demolition and excavation waste?

   a. It is buried on site

   b. It is burned

   c. It goes into landfill

   d. It is recycled

10. Which part of the Building Regulations 2010 requires the construction industry to consider and use energy efficiently?

    a. Part B

    b. Part D

    c. Part K

    d. Part L

Chapter 4

Unit CSA–L1Occ2
SETTING OUT FOR BASIC MASONRY STRUCTURES

LEARNING OUTCOMES

LO1: Know how to interpret information for setting out basic masonry structures

LO2/3: Know how to and be able to prepare resources for setting out basic masonry structures to the given instructions

LO4/5: Know how to and be able to contribute to setting out and constructing basic masonry structures to the given instructions

INTRODUCTION

The aims of this chapter are to:

* help you interpret and understand information

* help you to contribute to setting out basic masonry structures

* help you to construct basic masonry structures.

INTERPRETING INFORMATION

When you are setting out a masonry structure you will have to refer to a variety of paperwork. To begin with you will have to make assessments about any potential hazards. You will also need to refer to drawings and follow instructions. Each drawing will have been created using different scales. Mistakes can easily be made if you do not take account of the scales.

Hazards associated with setting out masonry structures

Setting out is one of the jobs that needs to be carried out at an early stage of any construction job. It refers to the marking out and positioning of a building. It is a very important operation, as it must be as accurate as possible. Mistakes made at the setting out stage are very costly to put right later; for example, incorrect measurements could mean digging out finished concrete. Buildings have sometimes had to be completely demolished after being built in the wrong place!

This is known as 'front end construction work'. Before setting out, services will have been diverted and any demolition work already carried out. Setting out will happen before any **ground work** begins.

It is worth remembering that the area where the new structure is to be built may contain some unpleasant surprises underground. For example, rubbish may have been buried there, so it is extremely important to make sure that you are careful when you clear the area ready for setting out.

In addition, there may be objects, such as cables, pipes, glass, metal and rubble just under the ground that may cause problems when you put the first pegs in. A site inspection should identify any potential problems.

When you are preparing the ground, make sure that you wear appropriate PPE, such as protective gloves.

KEY TERMS

Ground work

– this is preparation work, such as drainage and foundations. These are activities that must be undertaken before the rest of the construction can take place.

The following are documents you should refer to:

* A risk assessment is a document that identifies any possible risks or hazards in carrying out work such as setting out. It will explain exactly what tools, materials and equipment you will need, including PPE.

* Refer to a method statement – this will have been written by the site manager and will explain precisely what needs to be done and how the job needs to be carried out. It will take into account any risks or hazards identified in the risk assessment.

* Site safety rules – in addition to these two documents the site managers will be keen to ensure that all health and safety issues are considered. They may have ways of working that require even greater care than is legally required.

REED TIP

Make sure that it is YOU who wants to become a bricklayer. If you don't want to be one, you won't enjoy it. The people who are most passionate and motivated always turn up to work and college every day, arrive on time, and behave appropriately.

Information sources

In Chapter 2 we looked at how to interpret and produce information. To briefly remind you:

* Drawings and instructions – these might be rough sketches of the work, along with probably handwritten instructions. They will have been prepared by an experienced person.

* Symbols and hatchings – it is important to recognise the various symbols and hatchings that are used in construction drawings. These are like shorthand. Over time you will recognise them and understand exactly what they mean.

* Site safety rules – these are safety instructions and requirements. They will cover all of the regular jobs you are likely to undertake and have been written from experience and knowledge of the particular hazards that might crop up.

Reporting inaccuracies

A bricklayer is unlikely to be responsible for either surveying or setting out the project. However, if they are a supervisor, the responsibility falls on them to check that the work follows what has been outlined in the drawings. It means knowing if something has gone wrong or could go wrong.

Once the setting out has been done it is always good practice to double check. Additional checks should be carried out before important new stages of construction get underway. If there are any differences, the drawing needs to be checked to ensure it is correct. This means talking to the architect or designer to clear up any misunderstandings and avoid potential problems.

Scales

In Chapter 2, we looked at a range of different scales that are used to produce construction drawings. It is important to remember that if the drawing has a high second number in the scale then it is quite small compared to the actual masonry structure that is going to be built.

> **PRACTICAL TIP**
>
> Remember that the sizes of many of the buildings you will be constructing are very big. You should be able to use the scale to work out every single measurement from a scale drawing. These will be in proportion to the real thing.

Reading and taking measurements from drawings

It is important to be able to read and understand drawings and sketches. Some may appear complicated and you should take every opportunity to practise reading sample drawings.

In brickwork sketches are often drawn by an experienced bricklayer. They do this to show the size requirements of building components, or construction procedures.

Section drawings can be used to provide detailed construction of foundations or walls.

PREPARING RESOURCES

There is a lot of work and checking to be carried out before construction can begin. You will have to set up profiles and take various measurements. You will also need to record readings so that you can refer to them later.

This early stage of construction means that you can identify any potential problems and take action to put them right.

Risk assessment

In Chapter 1, we examined the importance of identifying potential hazards and the ways in which risk assessments and method statements can help to avoid possible health and safety hazards. It is also important to make sure that manufacturers' instructions are followed. All of these precautions ensure that you follow health and safety legislation and keep yourself safe.

Health and safety risks and PPE

The most effective way of handling health and safety on a construction site is to spot the hazards and deal with them before they can cause an accident or an injury. This begins with basic housekeeping and carrying out risk assessments. It also means having a procedure in place to report hazards so that they can be dealt with.

Work areas must always be kept clean and tidy. Sites that are messy, with materials, equipment, wires and other hazards can prove to be very dangerous. You should:

* always work in a tidy way

* never block fire exits or emergency escape routes

* never leave nails and screws scattered around

* ensure you clean and sweep up at the end of each working day

* not block walkways

* never overfill skips or bins

* never leave food waste on site.

PPE

PPE should include the following:

* Safety goggles or safety glasses – these are always necessary if there is a risk of injuring your eyes.

* Gloves – these are especially useful if you are working in cold conditions or where there is a risk of sharp objects sticking into your hands.

* Safety footwear – needed in case of sharp objects, uneven ground and poor weather conditions.

* High visibility (hi-vis) clothing – to ensure you can be seen easily on site.

Resources required for setting out

It may not always be necessary to have all resources available to you while you are setting out but Table 4.1 outlines what you might need.

Resource	Description and use
Site plan	A site plan is a location drawing that shows the position of the building in relation to the site and its surroundings.
Block plan	These identify the site in relation to the surrounding area.
Working drawings	These are prepared by the architect. They contain all the information that the builders will require to create the building as detailed by the architect.
Compass	To orientate the site a compass may be necessary to determine north. The site plan drawings will always show north and on site this assists in identifying the facing of the building.
Ranging lines	Ranging lines are stout lengths that are also known as shanks. For setting out four or five shanks are needed. They are usually made from hemp or nylon. Nylon can stretch more than hemp, but when used between pegs and profiles it will retain its tension better than hemp.
Builder's square	This is used to set out right angles on a site. A builder's square is wooden and braced with one side longer than the other. Squares are made from 75 mm × 30 mm timber and are half jointed at the 90° angle with a diagonal brace. Modern builder's squares are also often made from aluminium and can fold up.
Timber for pegs and profiles	Wooden pegs are used to establish corner points. They will also support the profile boards. Pegs will be needed to establish the level of concrete in foundation trenches and a peg will be needed as a datum peg. The pegs are usually 50 mm to 50 mm square. The profile boards are usually 100 mm × 25 mm.
Measuring tapes	Steel tapes are generally the most accurate and are available in lengths of between 3 m and 100 m.
Spirit level	This is a simple straight edge that has a glass tube containing a liquid and a bubble of air. When the air bubble is in the centre of the tube the straight edge is exactly level. The tubes are marked with lines to confirm that the edge is level.
Straight edge	This simple piece of wood or aluminium with parallel sides provides you with a straight edge, although not always with an entirely accurate one.
Hand tools (hammers and saws)	Primarily for preparing pegs and profiles, simple hand tools will be needed.
Optical level	An optical level has a mechanism to ensure that it is horizontal. It also has a telescope which magnifies the image.
Optical laser level	A laser level works on almost exactly the same principle as an optical level but only needs one person to operate it. A laser beam is sent out to a target receiver. The level at that point can then be recorded for setting out.

Table 4.1 Resources for setting out

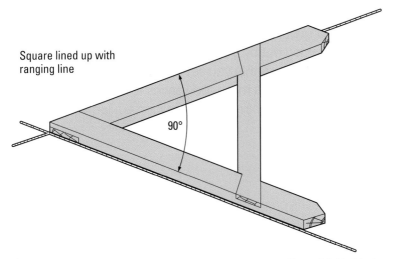

Square lined up with
ranging line

90°

Figure 4.1 Builder's square

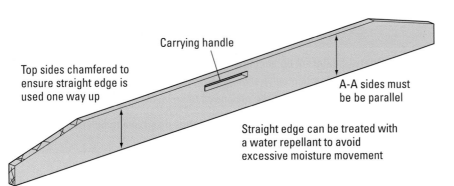

Carrying handle

Top sides chamfered to
ensure straight edge is
used one way up

A-A sides must
be be parallel

Straight edge can be treated with
a water repellant to avoid
excessive moisture movement

Figure 4.2 Purpose made straight edge

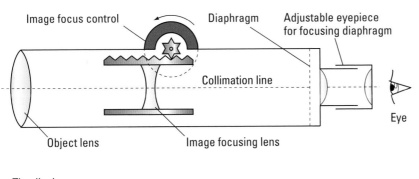

Image focus control

Diaphragm

Adjustable eyepiece
for focusing diaphragm

Collimation line

Eye

Object lens

Image focusing lens

The diaphragm
This positions the cross hairs

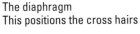

Cross
hairs

Stadia
lines

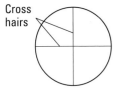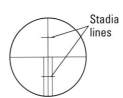

Figure 4.3 Diaphragm and telescope of an optical level

Checking resources

All equipment and instruments are limited in their use and over time they
may become less reliable. Relying on old, damaged or worn equipment
and instruments could cause major problems when setting out. It is

important always to test the equipment before using it, as instruments in particular tend to be rarely checked. They should have an annual calibration test carried out by a specialist and be certificated as accurate. You should always ensure that you set up instruments in the right way and if you find inaccuracies make allowances for them. Obviously, faulty instruments need to be repaired and damaged equipment replaced.

Site clearance

Site clearance means preparing the site for work. Each site will be different and some will need more clearance and work than others. This involves examining the site and identifying any potential problems, such as old foundations or rubbish that may need to be cleared away. Table 4.2 outlines example activities that may result from a walkover survey.

Site clearance issue	Explanation
Removal of obstacles	There can be a wide variety of different obstacles – remains of old buildings or debris will need to be removed before construction can get underway. There may also be underground obstacles that may need to be filled and the site may need to be levelled.
Hedges and treetops	Space is often at a premium, so hedges and low-lying branches may have to be cut back in order to provide the necessary space.

Table 4.2 Site clearance issues

Locating existing services

There can be a wide variety of different services, which could include:

* gas
* electricity
* water
* telephone
* electrical cable
* drainage.

The local authority should be able to provide you with information related to the location of services. However the locations still need to be confirmed by walking the site. Sometimes it may be necessary to use surveying equipment to find services that are hidden.

There are reasons for locating existing services before setting out:

* Setting out work will mean that excavation may be necessary. Services might have to be cut off or diverted. In any case, some pipes and cables may have to be exposed if trenches are being dug.

* Any exposed pipes or cables could cause a problem. They should only be exposed for the shortest period of time. Warning signs may be necessary.

* Eventually services will have to be reconnected. So it makes sense to figure out what connections will be needed later and to mark out where these cables and pipes will have to run.

SETTING OUT AND CONSTRUCTING BASIC MASONRY STRUCTURES

Setting out a masonry structure requires you to follow a series of key steps. These steps will ensure accuracy and correct positioning. For the setting out activity itself you will need to make sure that you have the correct equipment and materials:

* a steel tape measure, 30 m to 100 m long with 1 mm graduations

* timber pegs and profile boards

* ranging lines

* a builder's square and a straight edge

* spirit and optical levels

* a site square

* hammer and nails.

Correct location

The first task is to determine precisely where the construction is located. We have already seen the value of block and site plans. But it is also important to ensure that you orientate the setting out of the building by using a compass to determine north. You can then refer to the plans, as these will have north marked.

Building lines and dimensional accuracy

Each local authority's planning department will have a fixed building line. This is the line beyond which a building must not go past. The position of the building line is a given distance from:

* the centre of the road

* the kerb line

* boundary walls

* existing buildings.

The building line can affect the building itself in a number of different ways:

* A building can be on the building line.

* A building can be angled toward the building line.

* A building can be some distance behind the building line.

* Some parts of the building may be allowed to go beyond the building line, such as a porch.

The building line can be found in the block plan. It is usually shown on site by pegs identifying the corners of the front of the building.

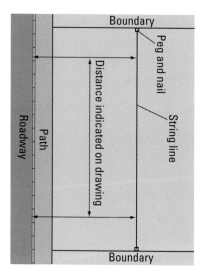

Figure 4.4 Building line and boundaries of site

You will need to mark the line by using pegs with nails on the top, with a ranging line stretched between each of the pegs.

When you are taking dimensions from a drawing always ensure that you check whether they are accurate. If something is unclear then report it to the site supervisor. There should be an accuracy of 1:1000, which means the measurements should not be more than 1 mm out for every 1 m.

The drawings will have been produced to a British Standard and will have been approved by the local authority, so they should be correct. However, use the dimension stated on the drawing, not the drawing itself as this can stretch during printing or photocopying.

Accuracy in setting out ensures that the correct dimensions are followed through to the finished construction.

Architects will normally have a disclaimer on the drawings which states that all dimensions must be checked on site and any discrepancies reported.

Setting out right-angled corners

There are several methods of setting out a right angled (90°) corner, some of which are shown in the practical tasks below. The 3:4:5 method of achieving right-angled corners uses a triangular shape to ensure that you produce a right angle. This can be seen in Fig 4.5.

The concept is straightforward. If you have a triangle that has sides with a ratio of 3:4:5 then a right angle will be created at A. The method is ideal for walls that are longer than 5 m.

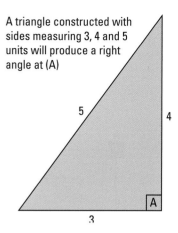

A triangle constructed with sides measuring 3, 4 and 5 units will produce a right angle at (A)

5

4

3

A

Figure 4.5 Right angles using the 3:4:5 method

PRACTICAL TASK

1. SET OUT A 90° CORNER USING A BUILDER'S SQUARE

OBJECTIVE

To make sure that all right-angle corners are square using a builder's square and checking it with the 3:4:5 method.

PPE

Ensure you select PPE appropriate to the job and site where you are working. Refer to the PPE section of Chapter 1.

TOOLS AND EQUIPMENT

Wooden pegs

Claw hammer

50 mm round nails

Two tape measures

Ranging line

STEP 1 Knock in two pegs (A and B) that are 1,500 mm apart. Knock nails into the centre of each peg, leaving enough of the nail sticking out so that you can attach the ranging line to them. Attach the ranging line. This indicates the face of the wall.

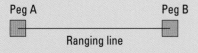

Peg A   Peg B

Ranging line

Figure 4.6 Setting out the face of a wall

STEP 2 Position the builder's square parallel with the ranging line that indicates the face of the wall.

PRACTICAL TIP

The builder's square must be accurately lined up with the ranging line. One way of doing this is to support the square by laying down some blocks and resting the square on them.

STEP 3 Adjust the ranging line by moving the end at peg B to peg C until it is parallel with the second leg of the builder's square.

STEP 4 Knock in peg C; knock a nail into the centre of the peg and attach the ranging line to it. Check again that the corner is square and make any minor adjustments if necessary.

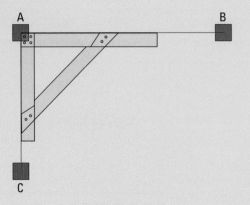

Figure 4.7 Using the builder's square

STEP 5 Now measure 900 mm along the ranging line from the corner or nail in the peg A at the face of the wall and mark it on the ranging line. This can be done by tying a piece of ranging line tightly onto the face line. Then recheck the measurement.

STEP 6 Measure 1,200 mm along the second leg of the corner and mark it on the ranging line as before.

STEP 7 Now measure between the two marks. The distance should be 1,500 mm if the corner is 90°. If it does not measure 1,500 mm then you will need to adjust peg C because you cannot adjust the face side.

You can also set out a corner using this method instead of a builder's square.

PRACTICAL TIP

It is important at this stage to check the accuracy of the corner. One way of doing this is with the 3:4:5 method, by taking three straight lines (one 300 mm long, one 400 mm long and the third 500 mm long) and joining them together to make a triangle. The angle opposite the longest side will be 90°.

You can use any lengths of line, as long as you keep the proportion 3:4:5. The longer the line, the more accurate the corner will be, as long as you use the same unit to multiply each side. For example:

Multiply 300 mm × 3 = 900 mm
400 mm × 3 = 1,200 mm
500 mm × 3 = 1,500 mm

PRACTICAL TASK

2. SET OUT A 90° CORNER USING AN OPTICAL SITE SQUARE

OBJECTIVE

To make sure that all right-angle corners are square using an optical site square.

PPE

Ensure you select PPE appropriate to the job and site where you are working. Refer to the PPE section of Chapter 1.

TOOLS AND EQUIPMENT

Three 50 mm × 50 mm timber pegs

Lump hammer

Optical site square and tripod

50 mm round nails

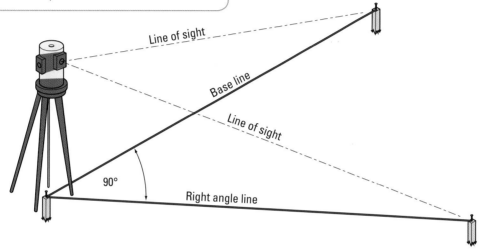

Figure 4.8 The site square in use

You can also make right angles using an optical site square, which contains two telescopes set permanently at 90° to each other. They can be adjusted vertically to enable you to look down at pegs that have been driven into the ground.

The site square is fixed to a tripod.

STEP 1 Secure the site square to the tripod.

STEP 2 Knock a nail into the top of the main corner peg and set the site square up over it. Site squares come with a steel rod or plumb bob (a line with a weight on it so that it will be plumb when hanging down). These are used to set the site square perfectly above the nail on the peg.

STEP 3 Using the bottom telescope of the site square, view along the building line to the other main corner peg. Adjust the site square until you can see the nail on the other peg in the centre of the cross hairs of the telescope.

STEP 4 Without moving the site square, look through the top telescope. This will give you a line at 90° to the building line. You can now knock in a peg into the ground in line with the view from the telescope and accurately knock a nail into the top of the peg in line with the cross hairs on the telescope.

REED TIP

When applying for a job, if there's space on your application form to write general comments, use as much of that space as you can to tell your future employer about yourself and your experiences.

Transferring levels

The term levelling means transferring levels to different positions on the site.

3. TRANSFER LEVELS USING A STRAIGHT EDGE AND SPIRIT LEVEL

OBJECTIVE

To transfer a given level to another point using a straight edge and spirit level.

All buildings need to be level so it is important to make sure that all the setting out is level at the beginning of the work. The simplest form of levelling is with a straight edge and spirit level, but this is normally only possible on smaller sites or over short distances.

This exercise requires points to be taken from a site datum. A site datum is a given position that all other datums on the site relate to. This can be items such as a kerb edge (with a nail inserted at the point), a window or door cill on an existing building, an inspection chamber cover or temporary wooden pegs knocked into the ground, known as a temporary bench mark.

PPE

Ensure you select PPE appropriate to the job and site where you are working. Refer to the PPE section of Chapter 1. **The photos for this and the tasks that follow were taken in a college environment so PPE was not deemed necessary. If you are working on a construction site you will need to wear PPE specified by your employer, such as hard hat and hi-vis jacket.**

TOOLS AND EQUIPMENT

Wooden pegs	Claw hammer
Straight edge	Lump hammer
Spirit level	

STEP 1 Knock a series of pegs into the area of ground that needs to be levelled. These pegs should be the same distance apart as the length of the straight edge. They will be level from the datum point or a set measurement above or below that datum point.

Figure 4.9 Series of pegs in uneven ground

STEP 2 From a given site datum, firmly knock the first peg (peg A) into the ground until it is approximately level with the site datum. Place the straight edge onto peg A and adjust it using the level until it is level with the datum.

STEP 3 Now reverse the straight edge and level (keeping the same edge on peg A), move the straight edge onto the next peg (B) and, using the level, level it with peg B.

Figure 4.10 Transferring levels using a straight edge and level

Figure 4.13 Reversing the straight edge (3)

STEP 4 Continue to level the rest of the pegs, reversing the straight edge as before.

Figure 4.14 Reversing the straight edge (4)

Figure 4.11 Reversing the straight edge (1)

Figure 4.12 Reversing the straight edge (2)

When levels have to be transferred over a long distance, using a straight edge and spirit level can be very time-consuming and rather inaccurate. A more accurate method is to use a quick set level but you will need another person to hold the staff while you use the quick set level.

PRACTICAL TASK

4. TRANSFER LEVELS USING A QUICK SET LEVEL

OBJECTIVE

To transfer a given level to another point using a quick set level.

PPE

Ensure you select PPE appropriate to the job and site where you are working. Refer to the PPE section of Chapter 1.

TOOLS AND EQUIPMENT

Wooden pegs

Quick set level

Staff

Claw hammer

Lump hammer

STEP 1 Securely position the tripod and attach the quickset level to it using the screw fitted to the top of the tripod. The tripod has spikes on the three legs and these need to be stood on and forced into the ground to stabilise it and keep it in a secure position

PRACTICAL TIP

It is important that the tripod does not move once the quick set level is placed onto it. If it moves, the instrument will not be accurate. If you set the tripod up on hard ground then put something against the legs of the tripod to stop it moving.

Figure 4.15 Quick set level

STEP 2 Adjust the instrument using the thumb screws at the base until the fish eye is accurately positioned. The fish eye is a level set into the base of the instrument. You position the telescope section over two of the thumb screws and adjust them until the fish eye is in the centre. Now turn the telescope through 90° and adjust the thumb screw until it is level that way as well.

Figure 4.16 Setting up a quickset level – screw thread of fish eye attaching level to tripod. Note the thumb screws

Figure 4.17 Quickset level showing the fish eye

PRACTICAL TIP

If you have to move, don't carry the tripod with the quick set level on it.

STEP 3 Get a second person to put the staff onto the datum peg and aim the instrument at it. Adjust the focus until you can read the staff.

Figure 4.18 Using a quick set level and staff

STEP 4 Adjust the cross hair on the view finder so that you can take a reading from the staff and write it down.

Figure 4.19 Taking measurements on a section of staff

PRACTICAL TIP

When reading from a levelling staff, you start at the bottom of the letter E. In the illustration you can see 2.9 m on the staff and each section of the letter E is 10 mm. Therefore the top of the E is 2.95 m. Again, each square above the E is 10 mm. If the cross hair is between two sections then you have to estimate the measurement. For example on the staff the reading between 2.92 and 2.93 is 2.923 m.

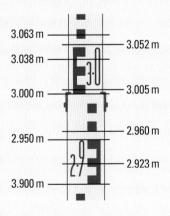

Figure 4.20 Measurements on a staff

STEP 5 Get the person helping you to move the staff to the next position to which the level is to be transferred.

STEP 6 The second person moves the staff up or down until the same reading appears in the cross hair of the quick set level as on the previous position. This is marked onto the peg or the peg is knocked down until it is level.

STEP 7 A timber cross rail can be nailed onto the peg so that the top of the cross rail will be in line with the transferred mark. This lets you rest the staff on it rather than trying to hold it against a pencil line, which could be rubbed or washed off if it rains.

STEP 8 Continue in this manner until you have transferred the datum level to all pegs.

PRACTICAL TIP

If you have to move the instrument, then you need to take a new reading from the last peg you levelled. If you are used to working alone then it would be better to use a laser level, which is set up just like a quick set level. Once the level is set up, it shoots out a red dot which you can position on the staff yourself.

CASE STUDY

South Tyneside Homes

South Tyneside Council's Housing Company

Taking your time will pay off

Gary Kirsop, Head of Property Services at South Tyneside Homes says:

'Setting out and building basic structures is all about making sure you take your time to set out correctly. This is so that by the time you come out of the ground, you're perfectly level. But again, while you're still learning, focus on developing the skills rather than finishing the job, so I'll say it again – take your time.

I've seen structures that have been built where they haven't taken their time, and you can see it's not level. Just with your eyes, you can tell it's not perfectly level where it's coming out of the ground – even a layperson can tell straight away – but especially someone who is working around buildings and construction all the time. When it's not level, it's unsightly.

So make sure when you're setting out your foundations of any structure, take your time and get the measurements right and the calculations right based on the size of the structure. You can see how important it is to do well in your GCSEs for maths. Of course it helps to be good with your hands if you want to be a bricklayer, but you also need pretty good academics too. At South Tyneside Homes, all our apprentices need to have achieved three or four GCSEs, including Maths and English.

Attention to detail is a critical skill in the construction industry. It matters to your employer, of course, it matters to the end client, and it should matter to you too. It's nice to stand back and take pride in your own hard work.'

Single walls and corner profiles

To keep ranging lines secure, profiles are used.

Profile boards are erected at each corner of the structure's footprint once the setting out is complete. They are used so that the corner setting out pegs can be removed to allow excavation work to take place. They are made from 50 mm × 50 mm timber pegs and 25 mm × 100 mm timber for profiles.

Normally the profiles are set at a datum level. This is usually the finished floor level (FFL) or damp-proof course level (DPCL).

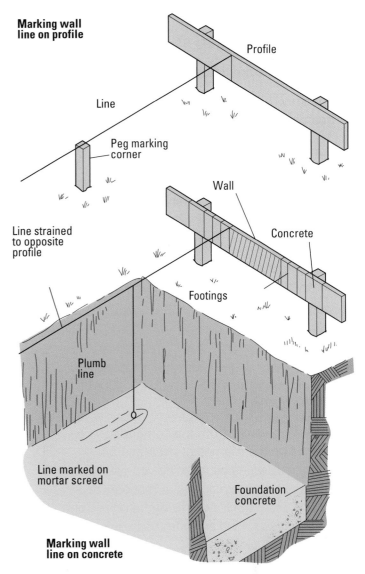

Figure 4.21 A marked out profile board

PRACTICAL TASK

5. ERECT CORNER PROFILES

OBJECTIVE

To erect corner profiles using a profile board.

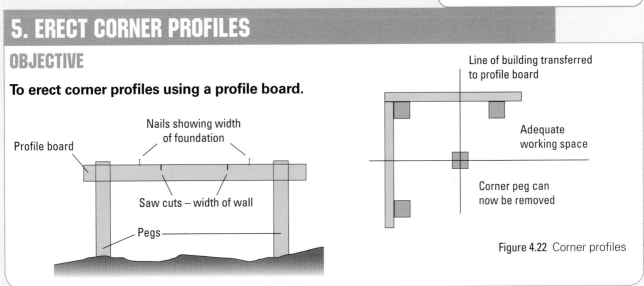

Figure 4.22 Corner profiles

TOOLS AND EQUIPMENT

50 mm × 50 mm timber pegs

25 mm × 100 mm timber for use as profile boards

Lump hammer	Spirit level
Hand saw	50 mm round head nails
Claw hammer	
Straight edge	75 mm round head nails or screws
Portable drill or screwdriver	

PPE

Ensure you select PPE appropriate to the job and site where you are working. Refer to the PPE section of Chapter 1.

STEP 1 Knock in two pegs at an appropriate distance away from the corner peg, parallel with the building line.

The distance away from the corner peg depends on the method of excavation. If you are excavating by machine you will need more room for the machine to manoeuvre, so approximately 1.5 m away from the corner peg. If you are excavating by hand then you can put the pegs a bit closer.

The distance between the two pegs should be the same as the length of the profile boards.

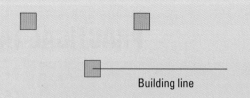

Building line

Figure 4.23 Two pegs running parallel with building line

STEP 2 Using one of the levelling methods previously completed, transfer the datum level to one of the pegs.

STEP 3 Using 50 mm round head nails or screws, attach the profile board to the pegs at the level marked on the peg. Make sure that the profile board is level between each peg.

STEP 4 Using a builder's square, knock in another peg at 90° to the other pegs. Refer to the practical exercise on page 118 if you need to refresh your memory on how to do this.

STEP 5 Using 50 mm round head nails or screws, attach the profile board to the pegs at the same level as the profile boards attached to the other peg. Make sure that the profile board is level between each peg.

STEP 6 Extend the ranging line fixed to the corner setting out peg and mark this line onto the profile board.

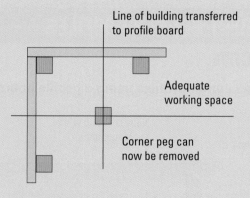

Line of building transferred to profile board

Adequate working space

Corner peg can now be removed

Figure 4.24 Transferring ranging lines to profile board

PRACTICAL TIP

It is not practical to just leave a pencil mark on the profile board as this could be worn away. At this stage you can use either a saw cut or a nail to mark the face of the wall onto the profile. If you saw cuts for the face of the wall, then use a nail for the edge of the foundation. Doing it this way, you will not get mixed up and think that the face of the wall is the foundation and vice versa.

STEP 7 You have now set out one corner. If you were setting out a building, you would re-check everything and complete the process for the remaining three corners. Once you are happy that all the dimensions are correct, you can remove the corner setting out pegs.

Walling and trench positions

Excavating foundations can begin once the ranging lines, which indicate the width of the concrete are in place.

The excavation of the trench can now begin. Once the foundation has been excavated, steel pegs are driven into the ground to indicate the depth of the concrete. The foundation concrete will come to the point at which the pins are fixed to the ground. This will ensure that you have the correct level.

Working space between profiles and excavation

Profiles should be approximately 1–1.5 m away from the trench edge. The actual trenches can be marked out using a spray paint, sand or dry cement.

There needs to be working space for the following reasons:

* If machines are digging the trenches then there needs to be room for the bucket arm.

* If they are being dug by hand there needs to be space for a wheelbarrow.

* If the profiles are moved to make room, then once the concrete is in the walls may not fit the walling.

Protecting setting out work

It is important that the profiles are not moved. They can easily become damaged and machinery could run over them. This would mean that all the measurements would have to be taken again. If damage or movement has not been noticed then there is a chance that the concrete will be put into the wrong position and the wall may not fit onto the foundations. This may be complicated and expensive to put right.

TEST YOURSELF

1. Who would usually write a method statement?

    a. Architect

    b. Designer

    c. Client

    d. Site manager

2. Which of the following can be described as a document that identifies the site in relation to its surrounding area?

    a. Block plan

    b. Site plan

    c. Working drawing

    d. Section drawing

3. Which of the following is a service that is likely to be found on a site?

    a. Gas

    b. Electricity

    c. Drainage

    d. All of these

4. Which of the following statements about building lines is **untrue**?

    a. A building can be on the building line

    b. A building can be angled toward the building line

    c. A building can be some distance behind the building line

    d. No parts of the building can go beyond the building line

5. What does the term levelling mean?

    a. Pulling down a building

    b. Transferring levels from the datum point to different positions on the site

    c. Setting a new datum point to match the foundation height of the building

    d. Flattening a kerb edge to match the foundation height

6. What is DPCL?

    a. Damp-proof course level

    b. Datum point clearance level

    c. Datum profile corner level

    d. Direct protective cleaning limits

7. When foundation concrete is put into a foundation trench, to what height should the concrete reach?

    a. Just above the first course of bricks

    b. Up to the pegs

    c. Just under the steel pins

    d. It doesn't matter

8. What is used to attach profile boards to pegs?

    a. Mastic

    b. Wood glue

    c. Round head nails/screws

    d. Staples

Chapter 5

Unit CSA–L1Occ30
MIX CONSTRUCTION MATERIALS

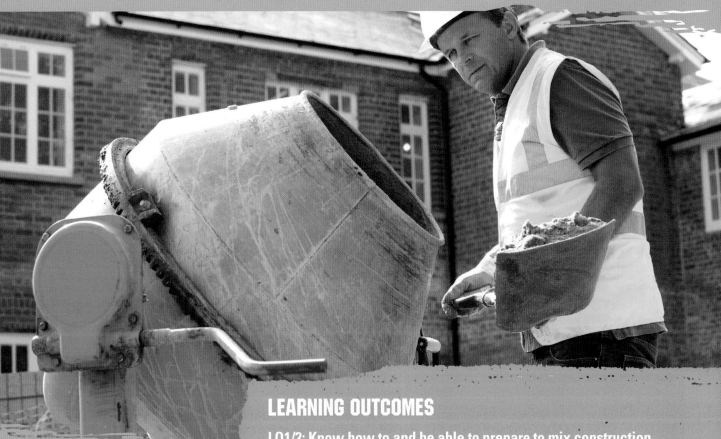

LEARNING OUTCOMES

LO1/2: Know how to and be able to prepare to mix construction materials

LO3/4: Know how to and be able to gauge and mix construction materials

LO5: Be able to store construction materials and components

LO6: Be able to restore the work area on completion of the work activities

INTRODUCTION

The aim of this chapter is to:

* help you to gauge and mix different types of construction materials.

PREPARING TO MIX CONSTRUCTION MATERIALS

Whether you are dealing with ready-mixed materials or mixing them on site it is vital that adequate safety precautions are taken.

The cement in concrete can burn the skin and is dangerous to the eyes. Skin contact with fresh, wet concrete, mortar or screed is often the cause of allergic or irritant dermatitis and it can also cause cement burns.

Above all, fresh, wet construction materials are very heavy. One cubic metre weighs 2.5 tonnes. This means it can cause strains and other injuries if not handled in the correct way.

Figure 5.1 Mixing concrete

Hazards and risk assessment

Each different type of construction material has its own potential hazards. Many of the materials can cause:

* irritation to the respiratory system if the dust is inhaled in the case of material such as dry cement

* irritation to the skin and particularly making the skin more sensitive or causing chemical burns in the case of wet cement

* damage to the eyes, leading to serious and potentially irreversible injuries in the case of wet cement.

Cement can cause problems whether it is wet or dry.

DID YOU KNOW?

Long-term skin contact with wet cement or fresh concrete can cause serious burns. It can also cause contact dermatitis.

Law	Key points
Control of Substances Hazardous to Health Regulations (1999) and Management of Health and Safety at Work Regulations (1999)	These require the employer to make an assessment of any health risks and then either prevent or control exposure to these risks.
The Construction (Health, Safety and Welfare) Regulations (1996)	All construction sites must have suitable and sufficient washing facilities, with hot and cold running water. They also need to have facilities so that you can change and dry your clothing.
Personal Protective Equipment at Work Regulations (1992)	Your employer has to provide you with suitable PPE. It needs to be properly maintained and replaced when needed. You should also be trained in how to use it.
Manual Handling Operations Regulations (1992)	Your employer should make arrangements so that whenever possible you do not have to manually handle materials or equipment. If necessary risk assessments should be carried out.

Table 5.1 Laws covering hazards when using construction materials

Information sources

There are four laws that relate to the hazards when using construction materials, including cement, mortar, plaster or concrete. These are briefly outlined in Table 5.1.

The type of information available about construction materials could include:

* Manufacturers' technical information – this information will tell you how to get the best out of the material, such as proportions to use and setting times. Importantly it will also include health and safety advice.

* Written and oral instructions – the work method statement should provide you with all the basic information that you need. Any additional specific or particular advice or guidance will be given to you verbally by experienced members of the team.

* Basic drawings and specifications – most construction jobs will have drawings and clear specifications stating the scope of the job and the materials that you should use.

* Health and safety legislation and official guidance – plays an important role in protecting you.

PRACTICAL TIP

The Health and Safety Executive produces a number of information sheets. Their Construction Information Sheet Number 26 looks at cement and highlights health effects and hazards. This can be found at www.hse.gov.uk/pubns/cis26.pdf.

Materials and components

Each particular job will require you to use particular types of material. For example, you may be required to mix mortar for bedding, jointing or pointing. This could mean that you would have to use ordinary Portland cement, sand and water. In addition you might need a plasticiser, a retarder or accelerator and, perhaps, a pigment or colouring agent. The actual mortar mix may depend on what it is going to be used for. If it contains a plasticiser or lime it will be more workable. If it contains cement it will stiffen up and set quicker. If it needs to be very strong it will contain cement, sand and water.

It is therefore very important to know what you are mixing the materials for and what properties are expected of them when they are mixed. This may mean that they need to:

* be workable for the maximum amount of time

* have sufficient strength to withstand weight once dry

* provide a good bond

* be durable to cold weather

* be sealed so they do not let in rainwater or draughts

* be of a colour that is complimentary to other building materials being used, such as bricks.

If you need to refresh your memory about the different materials and components then refer back to Chapter 3.

GAUGING AND MIXING CONSTRUCTION MATERIALS

One of the main materials that you will be using on site will be concrete. This is a mixture of cement, fine and coarse aggregates and water. Once they have been correctly mixed and poured into place the concrete will then set. As we will see, concrete is just one of the different types of materials that you will have to mix on a regular basis. The other common ones are mortar and plaster.

The mixes can be specified either by weight or volume. There are some different basic ways in which you can gauge and mix these materials to match your instructions. Each of them takes a different amount of time to prepare when you are hand mixing.

The other major concern will be how much time you have before the mix becomes unworkable. This means that it is important to only mix what is immediately needed, taking into consideration its setting time.

As with any work on a construction site, the first thing to do is to ensure that you have the right PPE and that you are working in a safe way.

PPE

When you are mixing construction materials it is important to remember the following:

* Never eat, drink or smoke when you are working with materials.

* When you have been working with dry materials you should get into the habit of washing or showering afterwards and using skin moisturiser.

* You should remove any contaminated clothing, including footwear and watches after you have been mixing materials. These should be cleaned before you wear them again.

* If you could be exposed to dust then you should wear some type of respiratory protection, such as a dust mask.

* The dust can also affect your eyes. If you are working with either dry or wet mixes wear approved glasses or safety goggles.

* For skin protection make sure you wear resistant gloves, boots and closed, long-sleeved protective clothing.

* If you are kneeling for any length of time you should wear kneepads, along with waterproof PPE if it is wet.

Correct use of PPE

PPE should be supplied to you by your employer. However, it is your responsibility to use it in the correct way. Make sure that you do not expose yourself to unnecessary dangers by ignoring PPE. Even if the job is only likely to take you a few minutes, failing to wear the right PPE could lead to an injury or a long-term physical problem. If you *should* be wearing protective clothing, dust masks, eye protection or any other form of PPE then make sure you *do* wear it.

It is also important to make sure that you store and maintain your PPE properly. If it is worn, broken or lost you should have it replaced.

In Chapter 1, we looked at the importance of PPE for most of the jobs on site. Refer back to this chapter now if you need to refresh your memory.

Safe working practices

You will be working with potentially hazardous materials. Just moving them around can lead to injury.

Some simple precautions, such as wearing the right PPE and being aware of what you are doing can go a long way towards making sure that you are working safely. It will make sure that you are protecting yourself and also others.

What is a hazard?

These are examples of situations that could become hazardous:

* If wet concrete or mortar falls into your boots or gloves or even soaks through your protective clothing you could be burned. The injury could take months to heal and you might need a skin graft.

* Inhaling dust created by moving or mixing materials can cause choking and problems with breathing.

* A clear working area should be set up so that you do not have to work around obstacles that could cause you to trip or slip.

Fires and fire extinguishers

In Chapter 1, we looked at fire hazards in the workplace. We also looked at fire and emergency procedures and the different types of fire extinguishers and their correct uses.

Recording accidents

Accident reporting and the laws related to it, along with the documentation, can be found in Chapter 1.

Dealing with emergencies

The correct way to deal with emergencies on site, records and the individuals involved was covered in Chapter 1.

REED TIP

Try to appreciate the different strengths and weaknesses of other members of your team. It will help you to work more effectively as a group.

Gauging and mixing materials

It is important to look at exactly how you gauge and mix concrete and mortars either by hand or using a mixer. It is likely that you will be using a range of different hand tools, mixing machines and equipment.

Each different type of mix has a particular set of ingredients. Gauging basically means proportioning materials. The mixes are given three numbers. For concrete this might mean:

* 1 – use one part cement

* 2 – use two parts fine aggregate or sand

* 4 – use four parts coarse aggregate or stone.

These are volumes or ratios and in this case 1:2:4.

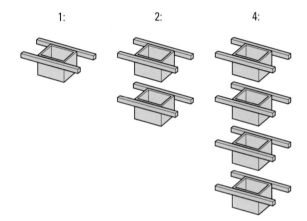

Figure 5.2 Mixing concrete by volume

Figure 5.3 One part of cement

Figure 5.4 Two parts of fine aggregate

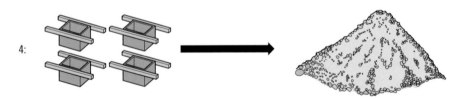

Figure 5.5 Four parts of coarse aggregate

Equipment used for mixing and how to use it

When mixing you can measure materials by volume or by weight. If you are measuring by volume you can use the following equipment:

* A shovel – this is bad practice as it is a crude way of gauging the amount of material you are using. This is because the amount of material on a shovel can vary. There is more sand in a shovelful than there is if you have a shovel of aggregate.

* A bucket – this is a better way because if the bucket is full it will have the same volume, whatever the material.

* A gauge box – this is a wooden box, which is made to size. The boxes do not have bottoms and once you have gauged the material, you can remove the box and shovel the material into the mix.

You can also measure by weight and there are tables that you can refer to that will show you how much material you will need by weight to create a cubic metre of concrete.

Mixing by weight is considered to be a slightly more accurate method. You can use a weight batch mixer. This records the weight of the materials as they are shovelled into the hopper. Then you just have to look at the calculations for the weights of the required mix and add the materials to the mixer. Caution: be careful if the material (e.g. sand) has been allowed to get wet as this will add to its weight.

Another alternative is to use an increasingly popular method known as the dry silo system. A stand-alone silo is delivered with ready-mixed materials to the site. They are still dry. Then it is only necessary to add the correct amount of water to the silo to produce the required amount of wet material.

You might mix the materials with a shovel. If you are mixing larger quantities, you might use a machine. Mixing construction materials using machines requires some additional equipment and materials. A good example would be mortar mixing, in which case you would need:

* the mixer – if it is petrol or diesel it will also need fuel. If it is electric then a waterproof cable connecting it to a generator or mains supply will be needed

* measuring equipment

* wheelbarrows, buckets and shovels to move the mortar to where it is needed

* the materials being mixed. For mortar this would be:

    * Portland cement

    * plasticiser

    * sand

    * water.

DID YOU KNOW?

Dry silo mortar systems hold around 33 tonnes of dry mortar. This is the equivalent of 23 cubic metres of mixed mortar.

All of this equipment and materials should be placed close to the working area so that the mortar can be mixed and quickly moved to the area as needed. This also means having all of the ingredients for the mortar mix close to where the mixing is taking place. A good example of this can be seen in Fig 5.6.

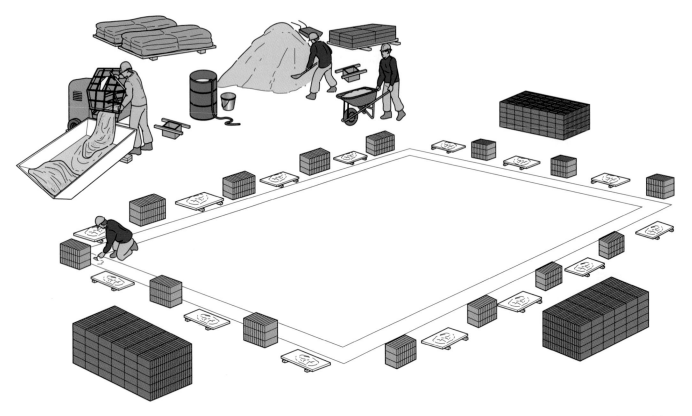

Figure 5.6 Placement of materials and equipment for mixing mortars by machine

Basic calculations

Calculating the exact mix you need to produce particular materials will depend on a number of things:

* the strength of the material you need

* what the material is being used for

* the specification for the job

* the required hardening time of the material.

Weight calculations

Table 5.2 shows you the weight of materials that are needed to produce different strengths of 1 cubic metre of concrete.

Strength in kg/cm²	Water/Cement Ratio	Cement kg	Fine Aggregate (Sand)	Coarse Aggregate (Stone)	Water (Litres)
350	0.52	355	880	950	185
300	0.58	320	910	950	186
250	0.66	280	940	950	185
200	0.74	245	970	950	182
150	0.86	210	1,000	950	181

* The table gives the total amount of water needed. However, if the sand and stone are very wet then the amount of water required will have to be reduced.

Table 5.2

In the table the first column shows the required strength of the concrete. This means that the concrete will have to support that load. The higher the number, the greater the load the concrete will have to support.

The other figures in the table relate to the ratios of materials needed. If, for example, the concrete strength was specified as being 250 kg/cm² then the following are required:

* 280 kg of cement

* 940 kg of sand

* 950 kg of stone

* 185 litres of water.

This would give you a water to cement ratio of 0.66.

Gauging mortar

When you are producing mortar it is important that the mortar mix is the same throughout the whole job. You want a consistent strength and look. This often means that measuring the mix using shovels is too inaccurate.

The way around this is to get the proportions right by weight, although you can use gauge boxes and buckets to get the right proportion by volume.

If mortar is proportioned by volume you will see a ratio in the specification, as can be seen in the Table 5.3.

Mortar type and ratio	Actual ingredients
Cement/lime mortar 1:1:6	One part cement, one part lime, six parts sand
Cement mortar 1:6	One part cement and six parts sand
Lime mortar 1:6	One part lime and six parts sand

Table 5.3 Mortar ratios

Note, however that lime is rarely used in modern bricklaying, though it may be used for specialist or heritage jobs. Note that hydrated lime is extremely alkaline, which means it will burn skin. Ensure you wear appropriate PPE when using it. Examples of the ingredients needed for each of these types of mortar can be seen in Fig 5.7, Fig 5.8 and Fig 5.9.

PRACTICAL TIP

When mixing by weight the amount of water required may actually be less if the sand or stone is very wet.

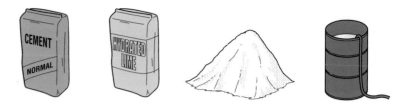

Figure 5.7 Cement/lime mortar – cement, lime, sand, water

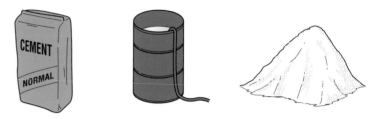

Figure 5.8 Cement mortar – cement, sand, water

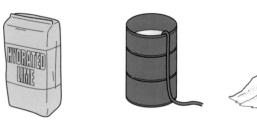

Figure 5.9 Lime mortar – lime, sand, water

Concrete

Concrete can be mixed either by hand or using a mixer. Again you will see a ratio mix, which, for example, might be 1:2:4. This means that for every one part of cement you will need two parts fine aggregate or sand and four parts coarse aggregate.

Figs 5.10–5.17 show how the process of mixing these ingredients works in practice.

Figure 5.10 Hand mixing concrete

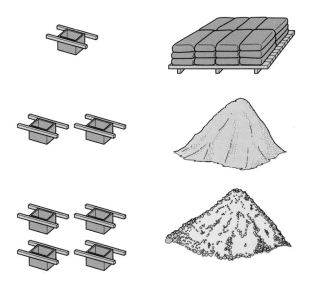

Figure 5.11 Materials for the mix

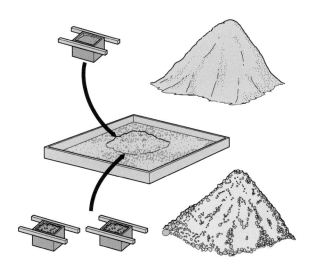

Figure 5.12 Mixing the sand (or fine aggregate) and coarse aggregate
– first place half of the sand or fine aggregate then half the coarse
aggregate

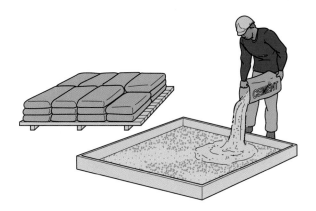

Figure 5.13 Placing the cement – place around half of the cement on top of
the sand (or fine aggregate) and coarse aggregate

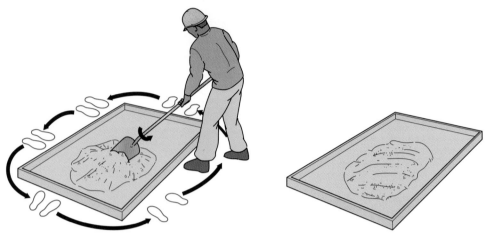

Figure 5.14 Turning over concrete – use a shovel to mix the concrete at least three times until you have a uniform colour

Figure 5.15 Making a hollow for the water

Figure 5.16 Adding water and mixing

Figure 5.17 A chopping motion is used to mix the materials

Characteristics, uses and limitation of materials

Remember the following:

* Mortar mixes – a 1:4 mix means that it is one part cement to four parts sand. This is the strongest mix and is used on engineering bricks. A 1:6 or 1:8 ratio is normal for bricks or blocks.

* Sand – this has grains that are smaller than 5mm and are classed as a fine aggregate. The sand needs to be clean and free of any organic matter.

* Course aggregates – this means gravel or crushed rock. These have grains that are larger than 5mm.

* Water – always use potable (drinking) water and never use water from ponds, water tanks or any other source. Only drinkable water will be free from pollutants and organic matter.

Possible defects

Construction defects cost around £20 billion per year in repairs and rebuilds. Although much of this cost is due to poor drawings, incorrect instructions and poor communication, some of it is because of defects in materials.

Should you come across hardened bags of cement or splits in cement bags then the material is probably not suitable for use. Sand and other aggregates that have become contaminated will not be suitable either. If these aggregates have soil or plants in them then once they are added to the mix these organic materials will rot and weaken the concrete.

You should always report any defects that you come across directly to your supervisor or site manager. They will be able to make a judgement as to what to do with the materials and whether or not they are still suitable for use.

Many defects can easily be avoided by:

* correct and secure storage on site (this means keeping bags of cement and other materials dry)

* checking for defects when the materials are delivered (rejecting torn or damaged bags or any materials with other obvious defects)

* only using freshly mixed materials.

Maximum time for use of mixed mortar, concrete and plaster

The maximum time to use mixed materials is very dependent on the weather conditions. In hot, drier weather the material is likely to have a shorter usage time.

DID YOU KNOW?

There are many different types of aggregate. A good example is ballast, which is a mix of fine and coarse aggregates. You can also use natural aggregates, such as gravel. Increasingly, recycled aggregates are being used. These are made up of crushed brickwork and concrete and are a useful way to recycle demolition waste.

Mortar tends to have a maximum usage time of around two hours.

Concrete is workable for around one-and-a-half hours, but in hotter conditions this could be as little as one hour.

Plaster usually sets within one-and-a-half hours but if it is cold then it can take as long as two hours.

Preparation times for hand-mixing

Mixing time will vary depending on the amount and type of materials that you are using. The most important thing to ensure is that all of the particles that make up the mix produce a consistent wet material.

Mortar can generally be mixed within two to three minutes in a mixing machine.

Each manufacturer will suggest their hand and machine mixing times. Multi-purpose concrete, for example, which includes a ready dry mix of cement, graded sand and aggregate mixed with water in a machine between three and five minutes. But this can take longer by hand, as you need to get the right workability and consistency.

Manual handling and lifting

It is important to remember that when you are moving construction materials around, particularly bags of cement, you should avoid carrying anything heavier than 20 kg. Try to use lifting equipment, such as wheelbarrows. Move larger quantities of material using pallet trucks or larger wheelbarrows.

In Chapter 1 there is lots of information and advice about manual handling.

REED TIP

If you're having trouble, always ask for help. It's better to ask and find out than to blunder on and make mistakes. By getting the help you need, you'll learn the skills properly, you'll avoid wasting time and money, and you'll have a better chance of succeeding in your future career.

Suitable working areas

Space can often be a problem on site. But whether you are mixing by hand or using a machine you need enough space to have all of the mixing equipment next to the materials. In addition to this you need to be close to where the materials are going to be used.

As we have already seen, if you place the materials and equipment close to where you are working this will provide you with quick access to the things you need so that you can work more efficiently. It will also mean that once they are mixed they have to be carried shorter distances.

When you are mixing and measuring by hand, you should also try to have the water and the materials close by, as can be seen in Fig 5.19.

Figure 5.18 Place the mixing equipment next to the materials

Figure 5.19 Measuring by hand

Protecting the work and its surrounding area

Many on-site activities can cause damage to the local environment. The effect can be even worse if waste materials are disposed of in the wrong way. For example, you should not hose down concrete slurry so that it goes into the main sewers.

The other issue is that you should protect the materials that you are using, ensuring you protect cement or mortar from the weather and interference while you are using them or if you need to leave them out. It also means that you do not leave piles of these materials around the site.

In each of Chapters 6, 7, and 8 that look at block walling, brick walling and cavity walling, there is useful advice about protecting the environment and the work that you are doing. You should turn to these chapters to see how it is relevant to the job you are doing.

Minimising damage

Freshly mixed materials should be protected from the elements also during use. The danger is that additional water from rain could weaken the mixture. It is also important to make sure that any mortar or cement that has been laid is protected until it has hardened properly.

The material has to cure before it is ready to do the job that it has been specified to do.

Maintaining a clean workspace

Mixing areas are likely to be used to produce a range of different construction materials. It is good practice to begin with a clean and clear area. In doing this you will make sure that whatever mix you are creating has not been polluted by other ingredients from a previous mix. It will also mean that any other polluting materials, such as organic matter like soil, will not get into the mix.

You must also make sure that any hand tools or mixing machines are clean. Tools such as shovels and wheelbarrows that are being used on a regular basis to mix materials will rust. Timber parts will rot, so they all need to be thoroughly cleaned and dried after each day's work. Mixing boards should be scrubbed down between each different mix.

It is also good practice to give trowels and other metal tools a coat of linseed oil from time to time to prevent them from going rusty.

Dust will get everywhere and this means that the exposed parts of any mixing machines will tend to accumulate cement dust. These should be cleaned on a regular basis.

PRACTICAL TIP

When you are cleaning and maintaining tools, always follow the manufacturer's instructions. Mixers should be serviced daily. This will ensure that they last longer and are easier to use. This may mean lubricating, cleaning with water and a stiff brush and checking oil levels.

DID YOU KNOW?

If you do not dispose of waste safely you can be fined an unlimited amount by the Environment Agency.

KEY TERMS

Noise bund

– this is an embankment that is often constructed around housing to cut out road noise.

Disposing of waste

The person or company running the construction site has what is known as a 'duty of care'. This means that they have to take responsible steps when they are dealing with waste. Therefore if they arrange for someone to come to collect the waste, they must check that this person is an authorised collector who will dispose of it properly.

Sustainability and recycling

In Chapter 3, we looked at sustainable construction. But even when you are working on the smallest project you need to be aware of the waste that you are creating. The UK government is keen to stop the huge amount of waste going into landfill sites. Many materials that are used in construction can be reused. The following are some examples:

* Aggregates, concrete, crushed bricks and stones can be used to build **noise bunds.**

* Rubble can be used as foundation material for roads, car parks and tracks.

* Soil can be used for landscaping on other sites.

* Wood waste can be used for footpaths, tracks and bridleways.

PRACTICAL TASK

1. MIX CONSTRUCTION MATERIALS

OBJECTIVE

To gauge and mix different types of construction materials.

MIXING MORTAR BY HAND ON SITE

Small quantities of mortar will often be mixed by manual means rather than by using a mechanical mixer.

A clean, hard surface should be used for mixing the materials. A concrete surface can be cleaned effectively after mixing but will stain if mixing occurs directly on it.

An alternative surface would be a large piece of plywood board which is free from debris or anything that may contaminate the mortar mix.

The materials used are gauged using a bucket, or a gauge box if one is available, and the materials are allowed to fall into a pile in the middle of the mixing area.

TOOLS AND EQUIPMENT

Plywood board	Gauge box
Shovel	Wheelbarrow
Bucket	

PPE

Ensure you select PPE appropriate to the job and site where you are working. Refer to the PPE section of Chapter 1.

PRACTICAL TIP

Warning
When water is added to cement it gets hot and can cause chemical burns so be very careful when mixing it.

STEP 1 Select a clean, hard surface if possible or use a large piece of plywood board for mixing on.

It is better to use a large piece of plywood board even if you have a smooth concrete surface, to prevent the mortar staining the concrete.

STEP 2 Measure out the correct amount of materials for the mortar mix. In this case you are going to make 1:6 (1 part cement to 6 parts sand) mortar mix.

Before the mortar can be mixed, the ingredients have to be measured in their correct proportions (see page 123).

STEP 3 Once you have accurately measured out the correct proportions of materials, using a shovel, repeatedly turn (three times) the ingredients into a pile making sure that all the materials are thoroughly mixed together.

STEP 4 Form a hole or indentation in the middle of the pile of dry mix.

STEP 5 Add water to the mix. Allow the water to soak into the mixed materials, and then add more water.

The water should be 'drinkable' and not contaminated.

Mix the materials with the water from the middle until all the water has been incorporated into the mix.

Adding more water to the mix will make the mortar easier to lay but it will reduce the strength by up to 30 per cent.

STEP 6 The materials are then turned repeatedly until the mix is a fatty workable mass.

The mix is now ready for use.

MIXING MORTAR BY MACHINE

Before using any machine you need to be familiar with the manufacturer's instructions where they apply with regard to:

* setting up the machine, leaving a safe working space around it, and ensuring it is level

* carrying out initial checks

* checking if any fuel tank is full (it is dangerous to fill petrol or diesel tanks on any machine when the engine is hot)

* whether it requires oil

* ensuring that any electrical connection is 110V or that a 110V transformer is available

* ensuring that the transformer is close to the 240V connection, reducing the amount of cable carrying 240V to a minimum

* ensuring the cable is carried overhead in a safe position, away from water, barrow runs, etc. and is fully unwound off the drum. If not, the cable can overheat and ignite, if the mixer is left running for long periods of time

* making sure there is no damage to cables, plugs or the transformer casing.

When you are satisfied that the mixing area is safe, follow the mixing sequence.

TOOLS AND EQUIPMENT

Plywood board	Wheelbarrow
Shovel	Mixer
Bucket	110V transformer
Gauge box	110V electrical lead

PPE

Ensure you select PPE appropriate to the job and site where you are working. Refer to the PPE section of Chapter 1.

STEP 1 Start up the mixer.

STEP 2 Add about three-quarters of the total mixing water required.

STEP 3 Add two-thirds of the sand required.

STEP 4 Add the total quantity of cement.

STEP 5 Add the remaining third of sand.

STEP 6 Add the remaining water until the correct consistency is achieved.

STEP 7 Allow a further mixing time of two to three minutes.

STEP 8 Tip the mixture into the wheelbarrow for transporting it to the work area.

PRACTICAL TIP

Mortar should be used within two hours of mixing. Small quantities of water can be added in hot weather to retain 'workability' but mortar should never be remixed after the initial set has commenced.

Concrete

Mixing concrete by hand or machine uses exactly the same methods but uses:

* cement (a binding material)

* sand (fine aggregate)

* gravel (course aggregate)

* water (should be drinkable grade).

STORING CONSTRUCTION MATERIALS AND COMPONENTS

Aggregates are often stored in bays, as can be seen in Fig 5.20.

Each different material or component needs a different method of storage. But this will all depend on how big the site is. It will also depend on how long you will be working on the site. If you are working on the site for a considerable amount of time then more permanent and effective storage should be organised.

This does leave a problem for small sites because space may be limited and proper facilities may not be easily available.

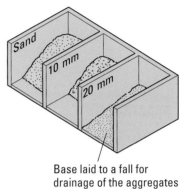

Base laid to a fall for drainage of the aggregates

Figure 5.20 Storage bays for aggregates

Correct storage

For sand and aggregates the use of storage bays is good practice on larger, long-term sites. For smaller sites, steel sheets and timber battens can be used.

On some larger sites purpose-built concrete bases are often constructed, with solid partition walls. Whichever method is being used the following points are good practice for the storage of aggregates:

* They should be kept clean and should be free of contaminants, such as debris and soil.

* They should not be allowed to get too wet, so storing them at a slight angle will encourage the water to drain away.

* Each different type of aggregate should be stored so it does not mix with another type of aggregate.

Storing cement presents another list of problems. On large sites, cement tends to be stored in large silos. This arrangement works very well. For smaller sites, the cement is delivered in bags and these need to be stored properly. For cement, good practice includes:

* making sure that the bags are stored in a well-ventilated and waterproof shelter

* making sure they are not stacked more than five bags high to prevent the weight of the bags compacting the ones below, so making the cement unusable. This is called warehouse setting.

* making sure they are stacked on pallets and not on the ground

* making sure the oldest bags of cement are used first and making sure that new and old bags are not mixed when deliveries are made.

Plaster should always be stored in a warm, dry place. Plaster will always attract moisture. If the plaster is exposed to damp and cold conditions it can set in the bag.

Damaged materials and components

In the previous section, we looked at ways in which you should report defects in materials. Remember that you need to routinely check the following:

* Dates on bags of cement or plaster. Always use the older bags first and do not use the bags that are out of date.

* Check any bags of cement or plaster, for example, which have been damaged or left open as they may not be suitable for use. Check to see if they have hardened or are damp. If they are, set them aside, as they should not be used.

* Make sure you are using the correct aggregates and that they have not been mixed with other materials. This will affect the mix that you are attempting to make.

* Also check materials have not become contaminated with soil or plant material.

RESTORING THE WORK AREA

As we have already seen, it makes life much easier when you begin the day with a clean and tidy work area. Getting into the habit of cleaning up after yourself at the end of the working day will mean that you can start work the following morning with a clean work area. This includes dealing with any waste that you may have created. Always make sure that the waste is disposed of in the correct way and that you do not unnecessarily pollute the environment.

Cleaning the mixer

The mixer must be cleaned thoroughly and carefully using plenty of water and a few shovelsful of gravel. Add the water first and then the gravel. Clean it out as soon as you can, as dried mortar is difficult and time-consuming to remove.

Do not attempt to clean the mixer using broken bricks as these may damage the mixer or at least reduce its working life.

Drain the water and gravel carefully, disposing of it according to the rules of your site. Don't just tip it out or down a drain.

Some sites use pressure washers to clean mixers – do not try to use one yourself without training.

Always immobilise the mixer at the end of a working day by disconnecting the power supply and/or storing the starting handle in a secure area.

TEST YOURSELF

1. Which law requires construction sites to have suitable and sufficient washing facilities?

   a. Control of Substances Hazardous to Health

   b. Personal Protective Equipment at Work Regulations

   c. Construction (Health, Safety and Welfare) Regulations

   d. Management of Health and Safety at Work Regulations

2. If a cement mix has a ratio of 1:3:6, how many parts of sand are required?

   a. 1

   b. 3

   c. None

   d. 6

3. If you are mixing concrete and the sand is wet, which of the following should you do?

   a. Reduce cement

   b. Add more sand

   c. Add more water

   d. Reduce the amount of water that you add

4. Lime mortar consists of which of the following?

   a. 1 part cement, 1 part lime, 6 parts sand

   b. 1 part cement, 6 parts sand

   c. 1 part lime, 6 parts sand

   d. 2 parts lime, 4 parts sand

5. If you see a ratio mix for concrete at 1:2:4, how many parts coarse aggregate are needed?

   a. 1

   b. 2

   c. 3

   d. 4

6. You have been asked to mix some concrete. Which of the following water sources should you use?

   a. Rainwater

   b. Pond water

   c. River water

   d. Drinking water

7. Under most normal conditions for how long is concrete workable?

   a. ½ hour

   b. 1 hour

   c. 1½ hours

   d. Most of the day

8. When storing cement on a large construction site, which of the following is good practice?

   a. Storing the bags outside under a plastic sheet

   b. Stacking the bags so that they take up the minimum amount of space

   c. Stacking the bags on the ground so they are stable

   d. Storing them in a waterproof shelter

9. What are the ingredients of ballast?

   a. Fine and coarse aggregates

   b. Sand and cement

   c. Crushed rubble

   d. Broken bricks

10. Which of the following is the least accurate way of gauging the amount of material you are using?

    a. Gauge box

    b. Shovel

    c. Bucket

    d. Unopened bag

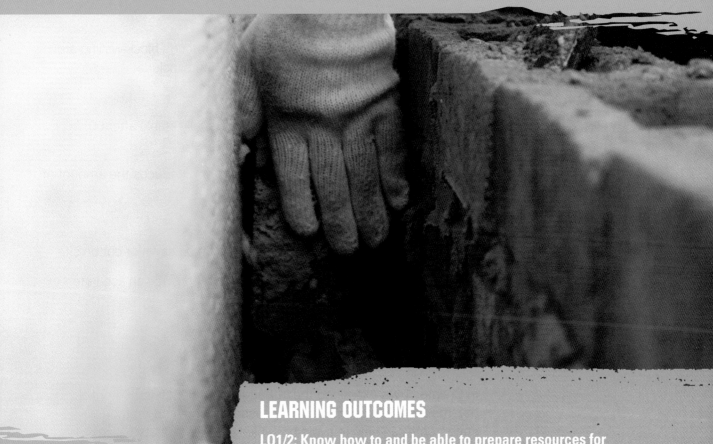

Chapter 6

Unit CSA–L1Occ19
CONSTRUCT BLOCK WALLING

LEARNING OUTCOMES

LO1/2: Know how to and be able to prepare resources for constructing block walling to the given instructions

LO3/4: Know how to and be able to set out block walling to the given instructions

LO5/6: Know how to and be able to construct straight block walling and return corners

INTRODUCTION

The aims of this chapter are to:

* help you to interpret and understand information

* help you to know how to set blockwork out to line

* show you how to construct block walls to the recommended height.

PREPARING RESOURCES

KEY TERMS

Aerated blocks

– these are lightweight blocks used for lightweight partitions and load-bearing internal walls.

The main resources you will use when constructing block walling are concrete blocks or **aerated blocks**. The benefits include:

* faster build speed

* increased productivity, as the mortar is usually stable after 60 minutes

* improved thermal performance, as blockwork reduces the amount of mortar being used by around 70 per cent

* good air-tightness

* reduced site waste, as often the blocks can be sawn or cut on site

* improved construction quality, as the blocks provide an accurate internal face that only needs a thin coat of plaster.

Potential hazards and risk assessment

In Chapter 1 we examined the importance of identifying potential hazards and the ways in which risk assessments and method statements can help to avoid possible health and safety hazards. It is also important to make sure that manufacturers' instructions are followed. All of these precautions help to ensure that you follow health and safety legislation and keep yourself safe.

It is important to remember that working on any construction can be hazardous. When you are constructing block walling it may also mean working at height. The health and safety legislation is designed to protect you. It aims to reduce the risks that you will face when carrying out your work.

Information sources

In Chapter 2 we looked at how to interpret and produce information. Here is a brief reminder:

* Sketches and instructions which will have been prepared by an experienced person.

* Symbols and hatchings – it is important to recognise the various symbols and hatchings that are used in construction drawings. These are like shorthand. Over time you will become familiar with them and understand exactly what they mean.

* Site safety rules are safety instructions and requirements. They will cover all of the regular jobs and have been written out of experience and knowledge of the particular hazards that might crop up.

Safety and PPE

The best way of handling health and safety on a construction site is to spot the hazards and deal with them before they can cause an accident or an injury. This begins with being tidy and organised and carrying out risk assessments. It also means having a procedure in place to report hazards so that they can be dealt with.

Work areas should always be clean and tidy. Sites that are messy with materials, equipment, wires and other hazards could be very dangerous. You should:

* always work in a tidy way

* never block fire exits or emergency escape routes

* never leave nails and screws scattered around

* ensure you clean and sweep up at the end of each working day

* not block walkways

* never overfill skips or bins

* never leave food waste on site.

There is some special advice about handling blocks and how to store them in the next section of this chapter.

PPE should include:

* safety goggles or safety glasses – always necessary if there is a risk of injuring your eyes

* gloves – especially useful if you are working in cold conditions or where there is a risk of sharp objects sticking into your hands

* safety footwear – needed in case of sharp objects, uneven ground and poor weather conditions

* high visibility (hi-vis) clothing – to ensure you can be seen easily.

Resources

You should produce a checklist for everything you might need and then select and check that your tools, equipment, materials and PPE are suitable for constructing block walling. This, of course, means making sure that any tools and equipment have been correctly maintained and are safe to use. This also applies to your PPE. It is also worth checking that your materials, including the blocks, meet the specification and are free from defects.

Tools and equipment

The range of tools and equipment required will depend on the site and on the job. The following list suggests the tools and equipment needed for a team of workers who are constructing blockwork:

* Laying tools – such as a brick trowel.

* Levelling tools – like spirit levels, straight-edge, gauge staff, line and pins, steel tape measure, square and bevel.

* Cutting and marking tools – such as a club hammer, bolster chisel, brick and comb hammer, cold chisel, brick cutting gauge.

* Finishing tools – like a pointing trowel, hawk, jointer and chariot.

Figure 6.1 Checking for plumb with a spirit level

Figure 6.2 A wheelbarrow can be a useful piece of equipment

The only other equipment that you really need is something to cut the blocks. A hammer, bolster and hand saw or a mechanical hand saw can be used.

Finally, you will also need a spirit level and a straight edge. This will help you to make sure that your blockwork is level.

DID YOU KNOW?

If you want to check to see if a spirit level is accurate then turn it over in a circle. If the bubble is still in the centre then it is accurate.

Materials

There are two main types of block – concrete or aerated lightweight blocks. Concrete blocks can be solid or hollow. Solid blocks tend to be used for commercial and industrial purposes. Hollow blocks often have rods running through them and they are then filled with concrete. Lightweight blocks have become commonplace to minimise lifting injuries when manually handling them.

Figure 6.3 Spirit levels

Aerated blocks are widely used for a broad range of different construction jobs. One of the main reasons for using these lightweight blocks is that walls can be built higher because the water from the mortar is absorbed into the blocks themselves. The walls are therefore more stable more quickly because the joints have dried faster than when using normal bricks.

Figure 6.4 Aerated building blocks piled ready for use

141

You will also need mortar. This is a mix of sand and cement with water. Several different types of mortar can be used:

* mortar with masonry cement

* mortar with ordinary Portland cement

* mortar with rapid hardening Portland cement.

The sand that you use for blockwork should be clean, soft and well-graded. This means as follows:

* It should not have any impurities – dirty sand will affect the quality of the mortar.

* It needs to be sharp rather than rounded – the best sand comes from quarries or pits and not from the sea or rivers (where the action of the water will have worn the sand down more).

* The grains also need to be of different sizes and not too fine or too coarse.

The final thing to note is the quality of the water used. It is always best to use drinking water from the mains, rather than using rainwater or standing water. If you use either of these then the water might contain impurities. This could also affect the strength of the mortar.

PPE

As we have already seen, it makes sense to ensure that you have your PPE handy and that it is in good condition.

You should have a basic range of PPE, including a safety helmet, hi-vis waistcoat or jacket, boots, gloves and eye protection. These should be provided by your employer and replaced if they get too worn or damaged.

Calculations

You will be working to a plan with a specification. This will tell you the types of materials that you are to use and the type of joint finish.

You will use the measurements on the plan to calculate the materials that you need to complete the job. It is good practice to list the different materials that you will need. This helps you to make sure that you have everything you need when you start the job so you will not be held up because you have forgotten something.

DID YOU KNOW?

You don't have to do all these calculations in your head – you can use a calculator.

Figure 6.5 Using a calculator may reduce the chance of a mistake

Standard blocks are 440 mm long and 215 mm high. As with any building material this can vary, but these figures are a useful guide.

When you are making calculations you also have to remember that there will be:

* a bedding joint of 10 mm

* a vertical joint, which is also known as a perp joint, between each of the blocks, of 10 mm.

This means that a pair of blocks has a length of 890 mm (440 + 440 + 10 mm) and a single block has a height of 225 mm (215 + 10 mm).

To work out how many blocks you need you have to work out the length and the height of the wall. As a rough rule, each square metre will need 10 blocks. This makes it easier to work out the total number of blocks required.

If, for example, you were building a wall that was 4 m long and 2 m high you would multiply the length by the height to give you the number of square metres. In this case it is 8 (4×2). We know that for each square metre we need 10 blocks so we would need 80 (8×10) blocks to complete the job.

If you are working with odd-sized walls, either in length or height, you might end up with a figure that suggests you would need a half or three-quarters of a block. A good example is a 2.5 m wall by 3.5 m. This gives us a total of 8.75 m² (2.5×3.5).

Using the guide of 10 blocks per square metre, it is suggested that you need 87.5 blocks (8.75×10). At the very least you should make sure that you have 88 blocks. If you also take into account the fact that some of the blocks may be damaged it is safer to have at least 90.

Figure 6.6 Cutting a block to size. NB: Gloves not shown for clarity but you should follow your college or employer's PPE requirements

Preparing and cutting blocks by hand

It may be necessary for you to cut and prepare blocks. A course of blockwork may not always be made up of whole blocks. To establish a solid wall with good bonds, blocks will need to be cut and prepared on site.

There are two ways of doing this by hand:

* Using a hammer and bolster chisel – this can be very fast and straightforward, but it is not a very accurate way of doing it. The cut does not have to be too neat, however, particularly if the blocks are going to be covered with a render or plaster.

* Using a hand saw – you can only use this technique for lightweight blocks. You have to mark the cutting line for a more accurate cut.

Protecting the work and the surrounding area

Newly built block walling can be vulnerable to cold temperatures and rain. The blockwork should be encouraged to dry out, which may mean putting a covering material over the face of a wall. Hessian is also used as an insulating layer.

Waste is unavoidable but strict environmental legislation determines how you handle and dispose of waste. The waste that is produced will cause some environmental damage and this waste must be minimised by:

* reusing broken bricks and blocks as hard-core

* sweeping fine debris into heaps and sprinkling water on it to minimise the dust

* not burning waste

* keeping all material bins shut or locked when not being used

* bagging up any waste, or at least covering it up.

Note that it is bad practice to return old mortar to be remixed.

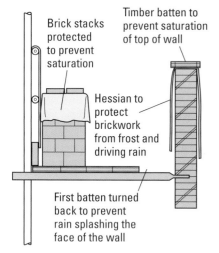

Brick stacks protected to prevent saturation

Timber batten to prevent saturation of top of wall

Hessian to protect brickwork from frost and driving rain

First batten turned back to prevent rain splashing the face of the wall

Figure 6.7 Protection of a wall

DID YOU KNOW?

Some parts of the UK have severe exposure to adverse weather conditions. These are areas that have higher than average rainfall, more frost or are over 90 m above sea level. Special precautions need to be taken in these areas.

GETTING READY TO CONSTRUCT

Once you get under way with building a blockwork wall you will be able to achieve quite a lot in a relatively short period of time. But all this depends on having everything ready and being well organised. Your progress will depend on which type of block you are using and what the weather conditions are like.

In this section of the chapter we look at all of the checks that you need to make before getting under way. We also look at the sequence of work and how to follow the standards required.

The practical tasks in this book assume that you have already done all the necessary preparation before starting to build. It is important to bear this in mind when completing these, or any other, construction activities.

Positioning blocks, mortar and components

Before starting the job it is advisable to stack all the components you need to work with.

It is very important that the working area should be prepared for working safely and efficiently before work starts.

The area should be levelled out, if necessary, to avoid tripping or stumbling, and should be cleared of all other building materials not required for the task.

Stacking materials
Be safe:

* Block stacks should be kept away from the edges of trenches.

* Walkways should also be clear and free for walking safely.

* Materials should not be stacked too high as there is a danger of the stack falling.

* The stack should be wider at the bottom than at the top to maintain stability.

Save effort:

* Bricklaying is a physical job and you need to make things as easy as possible for yourself. Setting out your work area is important.

* Usually the materials should be placed approximately 600 mm away from the wall you are building to avoid double-handling and wasting time and effort walking to and from the materials to the wall.

* Materials should be kept at a working height.

Protect materials:

* Materials should be protected against damage, dampness and frost.

* Ideally bricks and blocks should be covered with a tarpaulin or polythene sheeting to protect them from the elements.

* The stack should be straight and not too high.

PRACTICAL TASK

1. PREPARING THE WORK AREA

Before brickwork can be started, the working area must be stacked out with bricks and blocks.

It is not always a question of putting materials as close to the work as possible. Each building site may have different procedures. Some will encourage placing the components close by, but on others they may have to be kept in a compound until they are needed. You should make sure that:

* you follow any written instructions about the storage or placement of components

* if there are no written instructions you seek verbal instructions from the site manager

* you check with the site manager if you are concerned about where the components should be placed.

> **PPE**
>
> Ensure you select PPE appropriate to the job and site where you are working. Refer to the PPE section of Chapter 1.

STEP 1 Level the ground off for easy walking, and set up the mortar/spot boards about 600 mm from the face of the wall (or where the wall will be).

STEP 2 Space boards at the corners of the building and not more than 3 m apart along the wall length.

STEP 3 Stack bricks and blocks safely and within easy reach (600 mm away from wall).

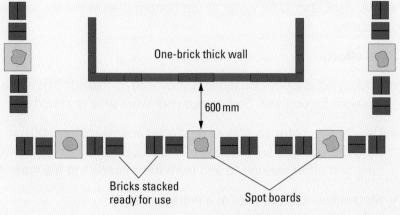

Figure 6.8 Setting out the work area

PRACTICAL TASK

2. ROLLING AND SPREADING MORTAR

OBJECTIVE

To get used to using a walling trowel to prepare mortar.

Rolling and spreading mortar is one of the most important arts of bricklaying – if you get this wrong you will not be able to lay bricks properly.

These instructions are for a right-handed bricklayer.

PPE

Ensure you select PPE appropriate to the job and site where you are working. Refer to the PPE section of Chapter 1.

TOOLS AND EQUIPMENT

Walling trowel

Mortar board

Mortar

METHOD 1: ROLLING MORTAR

STEP 1 Set up your mortar board and wet it before putting any mortar on it. Once you have wet your board you can put mortar on it but leave some space to roll the mortar around on the board.

Figure 6.9 Setting up the mortar board

Figure 6.10 Wetting the board

STEP 2 Using your walling trowel, pull away a small amount of mortar. With a sawing action, roll the mortar along the board until it looks like a sausage.

Figure 6.11 Pulling away a small amount of mortar

Figure 6.12 Rolling the mortar along the board

STEP 3 Now, using the edge of the trowel, slide it under the sausage of mortar and pick it up. It is now ready to spread the bed joint.

Figure 6.13 Using the edge of the trowel to slide under mortar

Figure 6.14 Picking up the mortar

PRACTICAL TIP

Some bricklayers do not like this method as you have to leave space on the board to roll the mortar. They prefer to have a board full of mortar. In this case you will see them use a different method as shown below.

METHOD 2: ROLLING MORTAR

STEP 1 Using your trowel, cut into the mortar, pick the mortar up and, turning your wrist, let it slide off the trowel. Repeat this two or three times until it looks like a pasty. It is now ready for spreading the joint.

Figure 6.15 Cutting into the mortar

Figure 6.16 Picking up the mortar

Figure 6.17 Pasty-shaped mortar

SPREADING MORTAR

> **STEP 1** Now turn to where you are going to lay your bricks and extend your arm. With the trowel just off the ground, swing your arm back towards yourself, turning your wrist, allowing the mortar to slide off the trowel.

Figure 6.19 Making a furrow in the mortar

> **STEP 2** Now, using the point of the trowel, make a furrow in the mortar. This makes it easier to lay the brick onto the joint. You might find that other bricklayers use a different technique but you must furrow a bed in some way to ensure it beds easily.

Figure 6.18 Laying out the mortar on the floor

Handling blocks

One of the main causes of injury when using blockwork is heavy loads. If you carry a load that is too heavy you can injure your muscles and tendons. HSE guidelines say that you should avoid lifting any heavy weights. If this is not possible, you should think carefully about how you will lift blocks or stacks of blocks to avoid hurting yourself. It is not simply just carrying the blocks that can cause problems. You may have to pass them up to others, or bend down to stack them.

If you are carrying heavy loads you are also more likely to slip, trip or fall. If someone else drops a block or a stack of blocks they could easily fall on you.

The Construction Industry Advisory Committee also points out that sharp edges of blocks can damage your skin.

The blocks are made from materials that can cause skin problems. The committee recommends the following health and safety points:

* Blocks should be stacked close to where they will be used.

* The block packs should be stacked on level ground.

* The block packs should not be double-stacked.

* Where possible, mechanical lifting and handling aids should be used (cranes, forklift trucks, trolleys etc.).

* Blockwork should be arranged so that you do not have to over-reach or twist when you are handling the blocks.

* The work should be arranged so that the blocks need to be handled only up to shoulder height.

> **DID YOU KNOW?**
>
> The Construction Industry Advisory Committee (CIAC) works very closely with the Health and Safety Executive (HSE).

Figure 6.20 Carrying blocks

Sequence of work and recommended block walling heights

On a building site you will have a drawing and measurements to help you plan your blockwork wall. Until you begin, you will not know just how many of the blocks you need to cut.

Normally this means having to set out the first course of blocks. If you do need to use a cut block to get the right length of wall it is normal practice to put it in the middle of the wall.

There are further complications, of course, if you have to create openings in walls.

Once you have the first course of blocks in position you will then see how your bond will work.

To establish a proper bond each new block should sit resting on two blocks beneath it. When you are starting at one end of a wall it may be necessary to begin the course with a half-block to establish the correct bond. The first half-block needs to be:

* gauge – this means that with the 10mm mortar joint it is at the right course height, in other words 225mm

* plumb – this means that the block is straight and is not too far forward or back

* level – this means that the wall as a whole is still level overall.

PRACTICAL TASK

3. BUILD A PYRAMID BY ROLLING AND SPREADING MORTAR

OBJECTIVE

To practise laying bricks or blocks level and straight with equal sized joints in order to achieve a sound wall and a good overall appearance.

Bonding is the arrangement of the bricks or blocks, usually overlapping between courses in order to:

* distribute any imposed load and to provide stability

* be decorative, as the bricks or blocks can be laid in a definite pattern to give a pleasing appearance and still maintain adequate strength for the job to be done.

To distribute any imposed loading, the bricks or blocks must be lapped over each other in successive courses both along the wall and across its width.

If the bricks or blocks are not lapped over each other then 'straight joints' occur; these are unsightly and considerably weaken the wall. Although bricks are used in this example, you can complete this task with bricks or blocks.

TOOLS AND EQUIPMENT

Walling trowel

Pointing trowel

Spirit level

Jointing iron

PPE

Ensure you select PPE appropriate to the job and site where you are working. Refer to the PPE section of Chapter 1.

PRACTICAL TIP

The frog is the indentation in the brick. Some bricklayers say that laying bricks frog up makes walls stronger than frog down but others say it makes no difference. You should learn both ways of laying the bricks as the architect or clerk of works may specify which they prefer for a particular job.

STEP 1 Set out the work area with the mortar boards and sufficient bricks to build a half-brick straight wall in stretcher bond.

Figure 6.21 Setting out the work area

STEP 2 Set out brickwork five bricks long in stretcher bond.

STEP 3 Lay the first brick and gauge it.

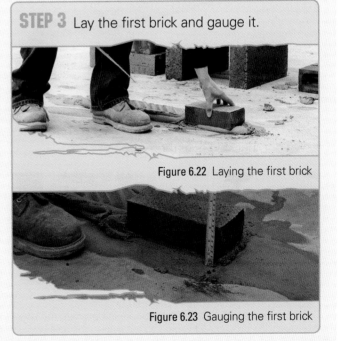

Figure 6.22 Laying the first brick

Figure 6.23 Gauging the first brick

Figure 6.24 Using the spirit level

STEP 4 Now lay the other end brick.

Level the other end brick with the first.

Figure 6.25 Laying the other end brick

Figure 6.26 Levelling the other end brick with the first

STEP 5 Lay the other bricks in between the two end bricks. Level them off.

Figure 6.27 Lay the other bricks in between the two end bricks

Figure 6.28 Level them off

STEP 6 Place a brick on the end as shown to give you a half bond. Complete the second and consecutive courses. Don't forget to gauge level and plumb each course.

Figure 6.29 Place a brick on the end

Figure 6.30 Gauge the level

Figure 6.31 The end result

The British Standard BS 5628, better known as the Code of Practice for the Use of Masonry, recommends that you should not build a block wall too high at any one time. They suggest that you should not build more than six courses of blocks. There are exceptions to this if you are building the blockwork as part of a cavity wall that incorporates wall ties.

> **PRACTICAL TIP**
>
> Six courses is only a recommendation. If the blocks are particularly wet it may be best not to build as many as six courses. It will take too long for the blocks to dry out and the water in the blocks will affect the mortar.

Industrial standards

Working to industrial standards means following clearly accepted guidelines and quality of work. These would include:

* erecting walls that are plumb (straight) levelled and finished

* ensuring that the wall is to gauge (the standard or required height and alignment)

* ensuring that the joints on all of the courses are plumb and true

* cleaning up the walling of splashes and excess cement or mortar when finished.

Safe working practices at height

As the height of the block wall increases it becomes impossible to carry out the work at ground level. This means that control measures need to be used to deal with any potential hazards. Working platforms are far safer than using ladders. Any working platform should have toe-boards and guard rails.

There is more information about working at height in Chapter 1.

Ensuring the work conforms to given instructions

It is important to carry out checks to make sure that the block walling is the correct length and height and is level, straight and plumb.

When you are dealing with lightweight blocks it is easy to knock them out of position accidentally. Blocks can also move on their own, particularly if they are wet. This process is known as swimming.

You should also check the materials that you are using against the specification for the task. You should never use poorer quality materials when specific materials are required as part of your instructions. As well as being important for safety and quality reasons this is something to bear in mind when you start working to clients' preferences. Imagine how you would feel if you paid someone to do a job on your house and they decided to change the materials without asking you.

> **KEY TERMS**
>
> **Swimming**
>
> – the block moves because of its weight and the presence of water. If the mortar is too watery then the block will slide or sink. If there is too much water in the block then this gets into the mortar and the block is even heavier.

It is common for some of the blocks to be broken or chipped. Minor damage may not present you with any particular problems, but you should not try to use badly damaged blocks.

Over time the tools that you are using will wear and eventually become unsuitable or unusable for the job. These need to be replaced. However a regular maintenance routine, including cleaning and correctly storing the tools and equipment, will increase their usable life.

The same is true for PPE. Not only should you have the basic PPE always available to use, but you should also report any losses or damage to your employer. It is the employer's responsibility to replace any damaged or worn PPE.

CASE STUDY

South Tyneside Homes

South Tyneside Council's Housing Company

Keeping it real

Gary Kirsop, Head of Property Services at South Tyneside Homes, started as an apprentice bricklayer 24 years ago.

'When you're doing basic blocklaying, it's important to take your time. It's not about speed; it's about understanding the importance of health and safety so you have a good grounding on live sites straight away. Make sure you do know the basics in college because the hazards on construction sites are real and bigger. Listen to your tutors and check that you're doing it right.

And when you're on your apprenticeship, listen to your mentors and team leaders; don't talk over them and say "I know better" – everybody is learning every day. Listening is a vital part of being a good communicator, and it is the best way to learn from tradespeople who are more experienced than you. If you're not sure how to do something, it is always better to go back and ask for help.

Think about the height that you're laying the blocks and their weight because there are different types. Make sure you're using scaffolds and trusses, and that you never lift above your ability. If you hurt your back, for example, think about the effect that could have on the rest of the job, your career or your life even. You also have to genuinely respect the limits of your workmates – not everyone will have the same physical strength.

Getting the right height is essential – there was an incident in the North-East not long ago… it was a windy day and some blockwork they were building got blown over because they were using thermalite blocks. So you can see how important it is to get the basics right. You have to understand the capabilities of each block and the maximum heights you should be building in any one day. So you have to think about how many courses you should be laying, and especially making sure your first couple are plumb and level.

Once you start coming onto live sites, the reality is that if it's not good enough, it has to be knocked down. That costs time and money. Remember that you'll be using real materials, working for real clients, real people, with real needs, who are spending real money.'

CONSTRUCTING BLOCK WALLING

Blockwork should be laid out at half bond. It is usual to lay out the blocks before bedding them into place, in order to look at the bond possibilities. Usually a cut piece of block will be needed to complete the course.

In this last section of the chapter we look at how you go about constructing block walling using practical models.

Constructing straight walls

The first practical task shows how you go about building a straight block wall. The practicals after that show how to build other features into a straight wall.

PRACTICAL TASK

4. BUILD A STRAIGHT BLOCK WALL BETWEEN PROFILES

OBJECTIVE

To get used to the process of laying blocks to a line.

This needs to be mastered before other skills, such as building corners.

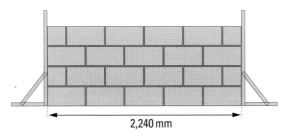

2,240 mm

Figure 6.32 You will build a wall like this one

TOOLS AND EQUIPMENT

Brick trowel

Lump hammer

Bolster chisel

Spirit level

Builder's square

Block/pins and line

Straight edge

Jointing iron

Steel tape measure

PPE

Ensure you select PPE appropriate to the job and site where you are working. Refer to the PPE section of Chapter 1.

STEP 1 Dry bond the first course and set up the profiles.

Figure 6.33 Setting up the profiles

Figure 6.34 Dry bonding the first course

STEP 2 If the floor is uneven, lay the first block at the highest end and transfer this to the other end using bedded blocks and a spirit level.

Figure 6.35 Gauging the first block

PRACTICAL TIP

When laying the beds, ensure that the mortar is a consistent width and height. The front of the bed should be flush and the back of the bed angled so that excess mortar does not squeeze out and drop to the back of the wall.

STEP 3 Gauge each course and run in.

Figure 6.36 Transferring the level

Figure 6.37 Gauging each course

Figure 6.39 Running in (2)

Figure 6.38 Running in (1)

PRACTICAL TIP

If the blocks are correctly placed, they will be level with the line as you look at the course horizontally, and set back 1–2 mm from the line as you look from above. Make sure the blocks do not touch the line.

PRACTICAL TIP

A good bed joint will mean that the block can be placed by hand pressure alone, although a lump or hammer can be used for adjustment. If you need to hammer the block down excessively, then the bed is too thick. Conversely, having to re-bed it because it is low to the line means that the beds need to be thicker. The only way to achieve this first time is by constant practice. A good furrow in the mortar will help with this process.

STEP 5 Cut the half-blocks by cutting round the entire block and over-hanging to assist the process.

Figure 6.40 Cutting the half-blocks

PRACTICAL TIP

When you are cutting a block, lay it on a stack of old blocks and let the waste piece of the block overhang the edge of the stack. This makes it easier and more accurate to cut the block.

STEP 6 Lay the half-blocks and then run the remaining courses into line, while making sure not to foul the line.

Figure 6.41 Laying the half-blocks

STEP 7 Complete each course until the wall is complete.

STEP 8 Finish the model with a flush joint.

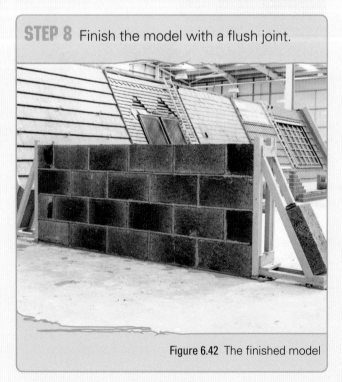

Figure 6.42 The finished model

PRACTICAL TASK

5. BUILD A STRAIGHT BLOCK WALL WITH PIERS

OBJECTIVE

To learn the methods of building a straight block wall with piers formed by blocks laid flat at either end.

This is a common way of strengthening block walls. As long as the two blocks laid flat are at the height of 225 mm, they will accurately set the corners for the remainder of the wall.

TOOLS AND EQUIPMENT

Brick trowel

Jointing iron

Lump hammer

Steel tape measure

Bolster chisel

Spirit level

Block/pins and line

PPE

Ensure you select PPE appropriate to the job and site where you are working. Refer to the PPE section of Chapter 1.

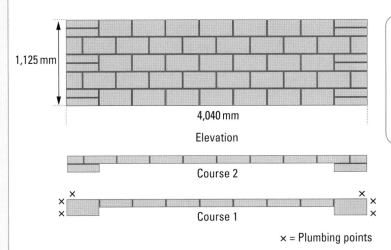

1,125 mm

4,040 mm

Elevation

Course 2

Course 1

× = Plumbing points

Figure 6.43 You will build a wall like this one

STEP 1 Mark a line to the length of the wall and dry bond the blocks along it to the required length.

Figure 6.44 Dry bonding the first course

STEP 2 At the highest end (if there is one), lay the first block flat and check for position and level.

PRACTICAL TIP

Lay the first block slightly higher than with a usual 10 mm joint as the next block has to reach the gauge height of 225 mm.

STEP 3 Lay a second block on top and check for plumb, level and to a gauge height of 225mm.

Figure 6.45 Checking gauge (1)

Figure 6.46 Checking gauge (2)

STEP 4 Repeat this process at the other end, while ensuring that the top block is level with the first end using a straight edge and spirit level to transfer the datum.

STEP 5 Check the wall for alignment by placing bricks on the two ends and fixing the corner blocks and line.

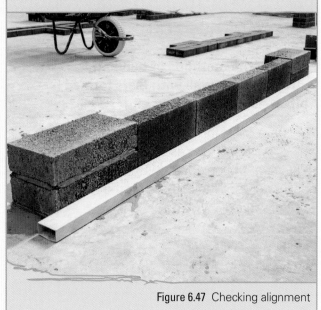

Figure 6.47 Checking alignment

STEP 6 Run in the first course and when completed make sure all the blocks are plumb.

Figure 6.48 Running in

PRACTICAL TIP

The line should be parallel to the end blocks and not kicking out or away from the line.

STEP 7 Build the corners up to full height, remembering to lay the blocks on edge as required.

Figure 6.49 Close up of the corners

STEP 8 Run in the remainder of the wall while maintaining a constant distance to the line and checking the bottom alignment of the blocks with the inner edge of your trowel as you cut the excess mortar away.

STEP 9 Finish with a flush joint to the front and rear of the wall.

Figure 6.50 The finished wall

PRACTICAL TIP

On the second course, the block at the back of the pier can have the edges 'clipped' (cut off) for the joints so that it can be placed more accurately.

6. BUILD A STRAIGHT BLOCK WALL WITH PIERS AND AN OPENING

OBJECTIVE

To learn how to construct a straight block wall that includes piers and an opening.

This model can be built by two students or reduced in size for a single trainee and should therefore be adjusted accordingly. If this model is constructed by one trainee then the length of the wall should be 2,690 mm (six blocks long).

PPE

Ensure you select PPE appropriate to the job and site where you are working. Refer to the PPE section of Chapter 1.

TOOLS AND EQUIPMENT

Brick trowel

Lump hammer

Bolster chisel

Spirit level

Block/pins and line

Straight edge

Jointing iron

Steel tape measure

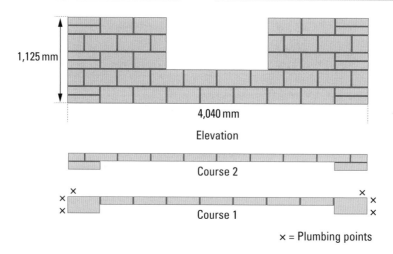

Figure 6.51 You will build a wall like this one

STEP 1 Mark a line to the length of the wall and dry bond the blocks along it to the required length.

STEP 2 At the highest end (if there is one), lay the first block flat and check for position and level.

PRACTICAL TIP

Lay the first block slightly higher than usual with a 10 mm joint as the next block has to reach the gauge height of 225 mm.

STEP 3 Lay a second block on top and check for plumb, level and to a gauge height of 225 mm.

STEP 4 Repeat this process at the other end, while ensuring that the top block is level with the first end using a straight edge and spirit level to transfer the datum.

STEP 5 Check the wall for alignment by placing blocks or bricks on the two ends and fixing corner blocks and line.

STEP 6 Run the first course in and, when completed, make sure all the blocks are plumb.

STEP 7 Build the corners up to full height, remembering to lay the blocks on edge as required. Make sure you use only the number of blocks specified at the corners as it is quicker to lay to the line. Also, make sure that the corners do not interfere with what will eventually form the opening.

STEP 8 Mark the position of the opening with lines cut into a screed of mortar and ensure that the blocks do not cross these markings.

STEP 9 Run in the rest of the wall while maintaining a constant distance to the line and checking the bottom alignment of the blocks with the inner edge of your trowel as you cut the excess mortar away. The edges of the opening should be checked for upright with a spirit level.

STEP 10 Finish with a flush joint to the front and rear of the wall.

Constructing corners

In this practical task you will look at how to construct corners in block walling. You will also look at the different ways you can establish bonds. It is important to remember that the half bond needs to be maintained, so when dealing with corners the blocks will have to be cut.

PRACTICAL TASK

7. BUILD A BLOCK RETURN

OBJECTIVE

To build a simple block return (corner).

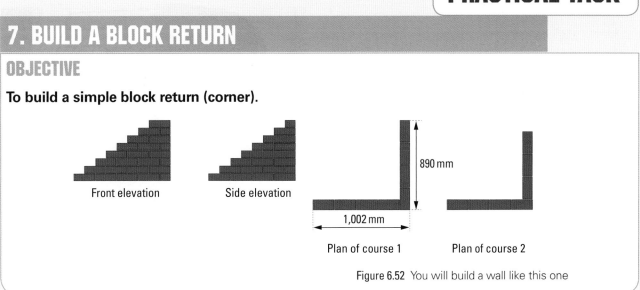

Front elevation   Side elevation   890 mm

1,002 mm

Plan of course 1   Plan of course 2

Figure 6.52 You will build a wall like this one

PPE

Ensure you select PPE appropriate to the job and site where you are working. Refer to the PPE section of Chapter 1.

PRACTICAL TIP

As blocks are taller than bricks, pay extra attention when plumbing them.

To make the blocks bond correctly, the introduction of a 100 mm cut piece in the right location will ensure that a half-lap bond is maintained.

TOOLS AND EQUIPMENT

Brick trowel	Builder's square
Lump hammer	Spirit level
Bolster chisel	Jointing iron
Steel tape measure	

STEP 1 Screed a 90° outline on the floor, making sure it is long enough for both sides of the return.

Figure 6.53 Screeding a 90° outline on the floor

PRACTICAL TIP

Scribe the outside corners of the markings so that the position can still be used if the lines are obscured by the mortar bed.

STEP 2 Lay the first block and check for gauge, level and plumb. The block also needs to be positioned correctly in relation to the position marked on the floor.

STEP 3 Lay the next block along and level to the first block while checking for plumb and alignment.

PRACTICAL TIP

The cross joint can either be spread on the block that is about to be laid or on the blocks that have already been placed. The choice is entirely down to the bricklayer's own preference.

STEP 4 Lay the remaining blocks along that side and check that these are 90° to the corner, adjusting accordingly.

Figure 6.54 Checking the return

STEP 5 Cut a 100mm piece from a whole block. This can be placed immediately next to the quoin (corner) block so that it is correctly placed to maintain bond.

Figure 6.55 Laying the 100mm piece

Figure 6.56 Checking for level

Figure 6.57 Dry bonding the first course

Figure 6.58 Positioning the blocks

STEP 6 Repeat these procedures with subsequent courses until the full height of the corner has been reached.

PRACTICAL TIP

Good, full width mortar beds will help prevent the blocks from leaning. If the blocks do start to lean; try compressing the beds with your trowel as this may be enough to maintain their position.

STEP 7 Finish with a flush joint on the front and back of the model.

Figure 6.59 The finished block return

8. BUILD A BLOCK WALL WITH A T-JUNCTION

OBJECTIVE

To build a section of straight block walling that incorporates a return (junction).

In this case the return is bonded into the main body of the wall. On site, indents could be left so that any additional sections can be built at a later date if required, allowing greater access and flexibility as work continues. However, this form of bonding requires the materials to be of the same densities and thermal properties to avoid the possibility of cracking. If the materials do not match, alternative methods of forming returns should be found, such as mesh or steel fixing methods.

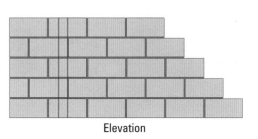

Elevation

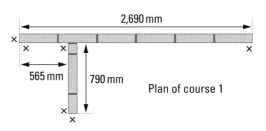

Plan of course 1

Figure 6.60 You will build a wall like this one

TOOLS AND EQUIPMENT

Brick trowel	Straight edge
Lump hammer	Jointing iron
Bolster chisel	Steel tape measure
Spirit level	Builder's square
Block/pins and line	

PPE

Ensure you select PPE appropriate to the job and site where you are working. Refer to the PPE section of Chapter 1.

STEP 1 Using the initial return corner, set out the return corner the other end with a two-block opening.

Figure 6.61 The second return corner

PRACTICAL TIP

The line should be parallel to the end blocks and not kicking out or away from the line.

STEP 2 Run the first course in and, when completed, make sure all the blocks are plumb. Place the blocks for the return and check for the correct dimension. Check for plumb level and check 90° to the main wall. At the other end, build a small corner and run in. Finish the model on both sides with a flush joint.

PRACTICAL TIP

On the return, start from the half end and drop the 100 mm piece in last. Then check the overall dimension is 890 mm to the far face of the main block wall. If this is achieved, then the two blocks on the second course will fit correctly.

Joint finishes

Jointing should never be left proud of the blocks. This means that the mortar should not stick out of the joint. The joint should be cleaned up as you work along the wall.

If blocks are going to be visible and the wall is not going to be rendered, then flush joints are usually recommended. A flush joint is a simple look. It means that the mortar is cleared off to the face of the blockwork. In other words it is in line with the edge of the blockwork. Normally this is done by using a piece of timber or a plastic block. Using either you can smooth off and compact the mortar into place.

Key points to remember

There are several basic points to remember before, during and after construction:

* When you are working in cold weather never dampen the blocks before you lay them. This could cause them to move around slightly as they dry out and the weather gets warmer.

* If you are constructing a wall that will need more than one pack of blocks you should select some blocks from each pack to build each course, rather than using up one pack then moving on to the next. This will reduce the risk of the wall having different bands of colour. There is no guarantee that each pack of blocks delivered will all come from the same original batch. This isn't relevant for interior walls or those that will be painted.

* Always use the right technique for cutting your blocks. On many sites there is a central cutting station. This ensures that all of the block cuts are even and standard.

* It may not be possible to build the entire piece of walling in a day or a few days. Poor weather might mean that you will need to leave it. In order to prevent damage to the wall and the joints you should always cover the work with a weatherproof sheet.

* While building the wall, make regular checks to see whether the length, height and finish of the wall looks accurate and similar.

* Slight differences in length or height can be easily dealt with by adjusting the bed or cross joint.

* If there are any particular problems with the completed work then you should contact the site supervisor and report it. It will be the responsibility of others to speak to whoever needs to be consulted to resolve a problem. You should not carry out any additional work until the problem has been resolved.

REED TIP

You'll be using your literacy and numeracy skills to keep accurate records, e.g. keeping track of how you use the stock you keep in your van so you don't run out.

TEST YOURSELF

1. Why might you use blockwork to build a wall?

   a. It is faster

   b. It has good air-tightness

   c. The construction quality would be improved

   d. All of these reasons

2. What is one of the items used to stabilise block walling?

   a. Props

   b. Gauges

   c. Wall ties

   d. Steel reinforcement

3. Where is sharp sand found?

   a. On the seabed

   b. On beaches

   c. In quarries

   d. In rivers

4. If you were a building a wall 6m × 3m approximately how many blocks would you need?

   a. 60

   b. 80

   c. 120

   d. 180

5. Which of the following is used to cut blocks by hand?

   a. Hammer and bolster chisel

   b. Mallet and chisel

   c. Edge of brick trowel

   d. Sledge hammer

6. If you have not been told otherwise, what distance should you place materials from where you are?

   a. 750mm

   b. 500mm

   c. 1,000mm

   d. 600mm

7. When you are working at height, which of the following is considered to be the most dangerous?

   a. Working platform

   b. Scaffolding

   c. Ladder

   d. Cherry picker

8. Which word is used to describe the measurement of each course of blocks to check they are at the right height?

   a. Plumb

   b. Gauge

   c. Level

   d. Standard

9. In normal weather conditions what is the maximum recommended number of block courses you should build at one time?

   a. 6

   b. 8

   c. 10

   d. 12

10. When you are storing packs of blocks before using them, how high should each stack of packs be?

    a. No more than five packs

    b. No more than three packs

    c. No more than two packs

    d. Only one pack

Chapter 7

Unit CSA–L10cc2
CONSTRUCT BRICK WALLING

LEARNING OUTCOMES

LO1/2: Know how to and be able to prepare resources for constructing brick walling to the given instructions

LO3/4: Know how to and be able to set out to construct brick walling to the given instructions

LO5/6: Know how to and be able to construct straight brick walling and return corners in half brick stretcher bond to the given instructions

LO7: Be able to construct one brick walling to the given instructions

LO8: Be able to form junctions in brick walls to the given specification

INTRODUCTION

The aims of this chapter are to:

* help you to interpret and understand information
* help you to set out solid brick walling to line
* help you to construct brick walls.

PREPARING RESOURCES FOR CONSTRUCTING BRICK WALLING

Many of the preparation tasks needed to build brick walling are the same as those we looked at in Chapter 6 and Chapter 2. You should refresh your memory on:

* potential hazards and risk assessments
* information sources and drawings
* safe methods of handling materials
* PPE
* preparing and cutting components
* protecting the work and the surrounding area from damage.

There are some pieces of advice that are relevant only to brickwork. These are covered under the next set of headings.

Potential hazards and risk assessment

Probably the most important hazards to consider in bricklaying are:

* the weight of the materials
* the fact that you are constructing walling that could be unstable until the mortar sets
* the fact that you will be working at height on some of the jobs.

The bricks weigh different amounts depending on their type. Clay bricks are the lightest but engineering or concrete bricks are much heavier. Just like blocks the bricks will also be heavier if they are wet.

As we will see later in this section, there are recommended ways in which to handle the bricks.

Information sources and interpreting drawings

Drawings are covered in detail in Chapter 2. For brickwork, drawings can be scaled using one of the following:

* 1:5

* 1:10

* 1:20

* 1:50.

Construction or working drawings and sketches are essential. These are diagrams that show both the building specifications and the procedures. In most cases the architect will tend to use **computer-aided design software**, or may draw the diagrams themselves. Sketches are simply that – they describe the particular requirements of either a construction procedure or a component.

The information that you are using should always be checked. Sometimes measurements on the drawings may not be accurate. So you might run into problems if the measurements on the drawings are wrong and they do not match the real site measurements.

Also, the documents may not have all the information that you need. Some information might be missing. The other problem may be that the information on one of the drawings might be different from that in another document. In all of these cases, if there is a problem you should refer it back to whoever prepared the drawings or documents.

Confirming instructions

It is important to check and confirm any instructions you are given. This will avoid misunderstandings. If you have not understood something but still go ahead then you could be wasting your own time and creating expensive mistakes if the work has to be done again.

You should always make sure that you fully understand what is expected of you. This goes for any written information, such as drawings or documents, as well as any spoken instructions you are given.

One very good way of making sure that you understood what is expected of you is to briefly repeat your instructions. Whoever is giving you the instructions will be happy with this approach. It will reassure them that you understand and that the work will be carried out properly.

The system works in the following way:

1. The person giving you the instruction should present it in the most useful way, whether this is a drawing, written instructions or a conversation.

2. That person will expect some sort of feedback. This could mean confirming that you understand or asking questions about something that you did not understand.

3. Once both of you are sure that you are in agreement and both understand then the communication has been successful.

4. You should not worry about going back to the person who gave you the instructions to ask them to confirm something that later on you were unsure about. At work people will always prefer you to check something rather than get it wrong!

Safe methods of handling bricks

When you are working with materials like bricks you will constantly be having to lift and stretch while carrying them. Unless you know how to handle them properly it is likely that you will injure your back, which is often a long-term problem.

You should never carry heavy objects alone – some people can carry more but others will find they can only cope safely with less.

Whenever possible you should try to use some kind of lifting equipment to do the hard work for you. You should avoid manual lifting whenever possible.

Bricks being transported up to a scaffold should be lifted with equipment like a telescopic handler.

Even if you are only carrying a small number of bricks you could still injure yourself. You should get into the habit of lifting things correctly. This means you should do the following:

Figure 7.1 Telescopic handler

* Before you start to lift the load just check to see whether the route you are planning to use is clear of hazards.

* Make sure that you have PPE, particularly gloves and reinforced boots.

* Bend your knees and keep your back straight, putting one foot slightly in front of the other.

* Keep your arms as close to your body as possible.

* Before you lift, check that what you are carrying is stable.

* Check the bricks have no sharp edges.

Figure 7.2 Brick trolley

* Get a firm grip on what you are lifting.

* Slowly lift and do not jerk the object from the ground.

* Do not twist your body if you need to put the load somewhere but change direction with your feet instead.

Other recommendations are:

* Bricks should be stacked close to where they will be used.

* Brick packs should be stacked on level ground.

* Brick packs should not be double-stacked.

* Brickwork should be arranged so that you do not have to over-reach or twist when you are handling the bricks.

* The work should be arranged so that the bricks only need to be handled up to shoulder height.

Sharp edges of bricks can damage your skin and they can made from materials that can cause skin problems.

Figure 7.3 Correct lifting posture using a brick tong

CASE STUDY

Going about it the right way

Marcus Chadwick has been a bricklayer with John Laing's then Laing O'Rourke for 18 years.

'These days there's more and more steelwork becoming involved; it's not just one brick on top of another anymore. We've got wind posts going in where we have to start drilling it into the metal work to fix all the ties. There's not that many corners to build now, because they're all junctioned by steel. Things have changed – I was still using a hod when I was an apprentice – a three-sided brick carrier with a long handle where you could stack 12 bricks, and you used to carry it on your shoulder. We're allowed to use them on the ground, but we're not allowed to use them on ladders anymore, because obviously you need your three points of contact at all times. Now we use brick clamps or tongs which we use to carry anything up to 12 bricks at a time in either hand.

Basically, you've got to maintain your perps. The first two courses that you lay are critical; that's your setting out point that dictates where your joints are going to fall. When you look down that wall, those perps have all got to be in perfect line. You get a bit of leniency now and again, but when we built Manchester Airport, that was measured with a 10mm piece of wood and an 1,800 level to make sure they were all in line. With the thickness of your perps and beds, don't be afraid to get your tape out and measure them. It's not going to make you look small or daft, it just makes you look more conscientious about your work.'

Calculating resources required

In order to work out the amount of materials required for a particular walling job, you need to know the area of the wall. The area is the surface that it covers. This is simply achieved by multiplying the length by the width to give the area of the walling.

For some jobs it can be more complicated, as there may be separate walls of different shapes. If this is the case you have to work out the area for each shape and then add them all together.

Working drawings will usually show lengths and widths in millimetres, so these need to be converted into square metres. It is often a good idea to convert from millimetres to metres before you start making any calculations.

If a working drawing shows 4,000 mm then you will need to divide it by 1,000. This is because there are 1,000 mm in each 1 m so 4,000 mm becomes 4 m.

Example

On a working drawing a wall is shown as being 9,500 mm long and 2,500 mm wide. The area of the wall in metres is therefore 9.5 m × 2.5 m = 23.75 m.

The formula to use is: L(length) × W(width) = A(area).

The next stage is to work out how many bricks you will need for each square metre of walling. Bricks come in a standard size so 60 bricks per m² is a good rule of thumb. But to complicate things, you may need to:

* take into consideration any openings, such as doors or windows

* add extra bricks to the quantity in case any of them are damaged

* add in additional bricks if you have one-brick walling or one-and-a-half brick walling. The calculation of 60 bricks per square metre is based on half-brick walling.

Once you know the total area of the wall that you will be constructing you multiply the number of square metres by 60.

Example

A customer wants a half-brick wall that is 6 m long and 3 m high. Assume that 5 per cent of the bricks will be damaged. So, 6 m (length) × 3 m (height) = 18 m².

$$18 \times 60 \text{ (standard number of bricks per m}^2) = 1,080$$

To take into consideration the 5 per cent wasted bricks, $(5 \div 100) \times 1,080 = 54$

$$1,080 + 54 = 1,134 \text{ bricks required.}$$

DID YOU KNOW?

Half-brick walling uses 60 bricks per square metre. One-brick walling uses 120 bricks and one-and-a-half-brick walling 180 bricks.

DID YOU KNOW?

For ordinary bricks, builders will often factor in a waste of around 3 per cent. For facing bricks this is 6 per cent.

Example

A customer wants a one-and-a-half brick wall that is 12 m long and 2.5 m high.

$$\text{So, } 12\,m \text{ (length)} \times 2.5\,m \text{ (height)} = 30\,m^2$$

$$30 \times 180 \text{ (standard number of bricks per } m^2 = 5{,}400$$

$$\text{Wastage: } (5 \div 100) \times 5{,}400 = 270$$

$$270 + 5{,}400 = 5{,}670 \text{ bricks required}$$

REED
TIP
...

A good apprentice is not just good with their hands, they're also good at Maths and English, they're reliable, and show a good attitude to work.

PPE

The Personal Protective Equipment at Work Regulations (1992) requires a risk assessment that will then highlight any PPE that will be needed.

PPE is there to help avoid or reduce risks. Depending on the type of job that you are doing, you will need the following:

* Some kind of eye protection – safety goggles will keep dust out of the eyes or safety glasses give basic protection.

* Hand protection – lightweight gloves will not give a great deal of protection. Plastic-coated gloves protect against chemicals. This means you may need to have two or three different types of glove to suit the particular task.

* Foot protection – it is always a good idea to have safety boots with steel toecaps and a steel mid-sole.

* Whole body protection – overalls or a high-visibility jacket are recommended. The overalls will give you protection from dirty conditions. In poor weather you might need waterproof or thermal clothing.

* Knee protection – if the walling job is at low height, particularly for the first few courses it may be a good idea to have knee pads.

* Hard hat – may prevent injury from items falling from height or stop you hitting your head on an object.

Resources required for constructing brick walling

Resources fall into three categories:

* Tools and equipment – which are dealt with in detail below.

* Materials – the bricks, mortar and any other materials that you will use as part of the build.

* PPE – this should be appropriate to the task and always be available.

Tools and equipment

All the tools that bricklayers use need to be:

* capable of withstanding a great deal of use and wear

* capable of giving many years of service

* used for the job for which they are intended.

Each of the tools required for walling is briefly described in Table 7.1.

Tool	Description
Bricklayer's trowel and pointing trowel	Trowels are made in sizes from 225 to 350 mm. The 250 to 275 mm blade is ideal for general work. There are left and right-handed blades. Pointing trowels range from 75 to 150 mm. Smaller trowels are known as dotters and larger ones known as bed jointers.
Spirit level	These are available in a variety of different lengths up to 1.8 m or more. Smaller levels are ideal for adjusting individual bricks. Longer ones are good for plumbing courses.
Brick hammer	These can have a hammer at one end and a blade at the other. Some will have a slotted end into which either a blade or a comb can be inserted.
Jointing iron	These are usually 60 to 125 mm long and used to form recessed joints.
Brick bolster and club hammer	Bolsters are used with club or lump hammers to cut bricks. Club hammers are around 1kg and usually heavy enough for cutting away holes in brickwork.
Line and pins	These are made of steel and bought in pairs. The line is bought in knots. For general purposes two knots of line are sufficient on one pair of pins.
Brick tong	These allow you to carry several bricks, usually 6 to 10, at one time in a safer way than trying to balance them. You lift a clamp and jaws clamp around the bricks. They are usually metal and fully adjustable.

Table 7.1 Tools required for walling

REED TIP

Cheap tools are not a good buy. It is worth investing in tools made by well-known manufacturers that have a good reputation. On many sites you will be judged by the quality of your tools and their appearance.

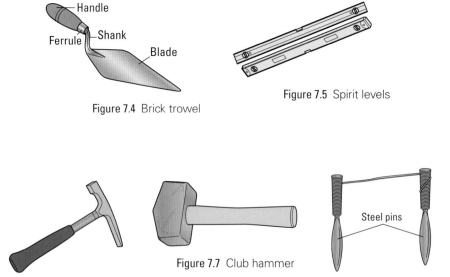

Figure 7.4 Brick trowel

Figure 7.5 Spirit levels

Figure 7.6 Brick hammer

Figure 7.7 Club hammer

Figure 7.8 Line and pins

Materials

The specification for the job will indicate which type of bricks you are required to use. Table 7.2 describes the four main types of brick and where they are usually used.

Type of brick	Description and use
Facing	The term 'facing brick 'can cover a wide range of different types of brick. These types of bricks are the most varied. Some of them are not that suitable for exposure to extreme weather conditions. They can be used for both internal and external work. Usually the bricks will be visible in the final structure. The most common is the red facing brick. It is available in a variety of different colours and can be soft sand-faced or hard with a relatively smooth surface. Mottled pink bricks are known as Flettons. These are used particularly for internal work. They have a variety of different surfaces. The final type of facing brick is known as a stock. These are machine-made now although they used to be hand-made. They are usually red in colour and can have a variety of textures.
Common	These are used for most standard brickwork that will be hidden either by plaster or some other covering. They can be wire cut or poorer quality stock bricks.
Engineering	These are hard burnt and dense bricks. They can be used for retaining walls, damp-proof courses and freestanding walls. They are particularly chosen for the job because they are capable of withstanding a good deal of compression and they have low water absorption qualities.
Special	These are bricks that are of a particular shape or size. They are created to either match facing bricks or to contrast with them. Special shaped bricks, such as squints, are designed so that you can build angled corners with them. This means that it is not necessary to have sharp corners and the thickness of the wall does not have to be increased.

Table 7.2 The main types of bricks and their uses

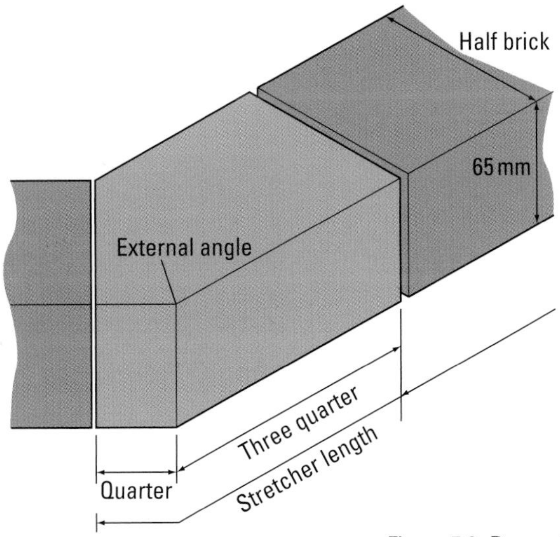

Figure 7.9 Proportion of a squint brick

Half brick

65 mm

External angle

Three quarter

Stretcher length

Quarter

KEY TERMS

Wire cut

– the bricks are made using clay. A large slab of clay is placed on a steel table and a frame with wires is brought down onto the clay, cutting it into a number of bricks.

In Chapter 5 we looked in detail at mixing construction materials, including mortar mixes. It is the mortar that holds the bricks together. The bricks are arranged into the mortar in various types of bond.

To refresh your memory about the types of materials used for construction you should look at the information in Chapter 3.

Figure 7.10 A Fletton brick wall

177

Preparing and cutting bricks by hand

It is sometimes necessary to cut bricks, but they need to be cut in a clean and accurate way. The method you use will depend on the type of brick being used. It needs greater effort to cut a dense engineering brick than an ordinary soft clay brick.

You can use a masonry saw to accurately cut ordinary bricks but in the majority of cases you should use a hammer and a sharp bolster. Many bricklayers will actually use their trowel or a brick hammer, but this is not recommended if you are working on face bricks.

Protecting the work and the surrounding area

Newly built brickwork can be vulnerable to cold temperatures and rain. The brickwork should be encouraged to dry out, which may mean putting a covering material over the face of a wall. Hessian is also used as an insulating layer.

Waste is unavoidable but strict environmental legislation determines how you handle and dispose of waste. The waste that is produced will cause some environmental damage and this waste must be minimised by:

* reusing broken bricks and blocks as hardcore

* sweeping fine debris into heaps and sprinkling water on it to minimise the dust

* not burning waste

* keeping all material bins shut or locked when not being used

* bagging up any waste, or at least covering it up.

Note that it is bad practice to return old mortar to be remixed.

Figure 7.11 Cutting bricks by hand

CASE STUDY

LAING O'ROURKE

Keeping it neat keeps it safe

Marcus Chadwick has been a bricklayer with John Laing's then Laing O'Rourke for 18 years.

'Get your space sorted, safe and clear and in order, before you even start. If you don't, you just end up falling over yourself and everything gets in the way. I work in a two and one gang – so we've got two bricklayers and one labourer. Our labourer is the best thing since sliced bread – he knows exactly where to put your pack of bricks, where to put your spot board for your mortar, he knows how to load the scaffold out on all its strongest points. It's important because once you've got your walls set out, you don't want things too close to you. You don't want to be falling over trailing leads on the floor from your drills and your guns. Make sure your housekeeping's good as well. Don't leave your debris lying about on your scaffold or your ground – make sure your ground's level, because slips, trips and falls are one of the main causes of injuries in the construction industry.'

SETTING OUT TO CONSTRUCT BRICK WALLING

Before you start any construction it is important to have all the resources ready and to know the order in which you should tackle the work.

Setting out resources ready for use

The advice on setting out your resources for Chapter 6 also applies to brick walling.

Sequence of work and recommended brick walling heights

In order to build any wall the first thing you have to do is to make sure that the corners are built in the correct position. This is often done by using corner profiles.

Corner profiles are right-angled steel plates that are removed after the brickwork has been completed. Once the corners are accurate a line is then attached. This is level to the course height. The line is pulled tight and this will give you the face of the wall.

It is now possible to begin laying the bricks at the correct height and in a straight line. The bricks are tapped into place so that they are level with the top of the string line.

It is important to get this first line correct. There should be a small gap between the line and the face of the brick. The gap is usually around the thickness of a trowel blade. The gap must be even along the length of the brick. This means that the brick should never touch the line.

It is recommended that the maximum number of brickwork courses to lay in one day is 16.

Bonding

When you are building walls with bricks it is important to make sure that they are laid out in a particular pattern. This is known as a bond. There are a number of different types of bond, but all of them are designed so that none of the vertical joints of a course is directly above another vertical joint. This is relevant on either a previous course of bricks or the next course up.

A bonded wall face with no vertical straight joints

Figure 7.12 A bonded wall

Bonding has three main purposes. These are to make sure that:

* the wall is strong and that any loads on it are evenly distributed

* the wall has lateral stability and that it is resistant to any side thrusts

* the wall is attractive and even.

Bricks need to be laid out so that they have maximum strength. They can be laid to half bond, which means that the brickwork overlaps half of the brick below it. The minimum overlapping should be a quarter bond (the brickwork overlaps a quarter of the brick below it).

If the brickwork is an independent wall then the second course of bricks should begin with a half-brick.

In order to understand bonding it is important to appreciate some basics about bricks. These can be seen in Table 7.3.

Term	Explanation	
Work size and brick terms	The standard brick is 215 × 102.5 × 65 mm. In a wall, if the 102.5 × 65 mm face is exposed then the brick is called a **header**. If the 215 × 65 mm face is exposed it is a **stretcher**. The surface underneath the brick is the **bed**. The depression at the top of the brick is a **frog**. The sharp edge is called the **arris**.	Figure 7.13 Brick with terminology
Coordinating size	Given the fact that mortar joints are added to each of the dimensions of the bricks, the actual coordinating size is larger than a standard brick size. Usually a 10 mm joint is added. This means that the coordinating size is 225 × 112.5 × 75 mm.	Figure 7.14 Work and coordinating size

Sizes

	Length	Width	Height
Co-ordinating	225 mm	112.5 mm	75 mm
Work	215 mm	102.5 mm	65 mm

Term	Explanation	
Queen closers	These are cut bricks. They are visible on the face of the wall. The queen closer is 46.25 × 65 × 215 mm.	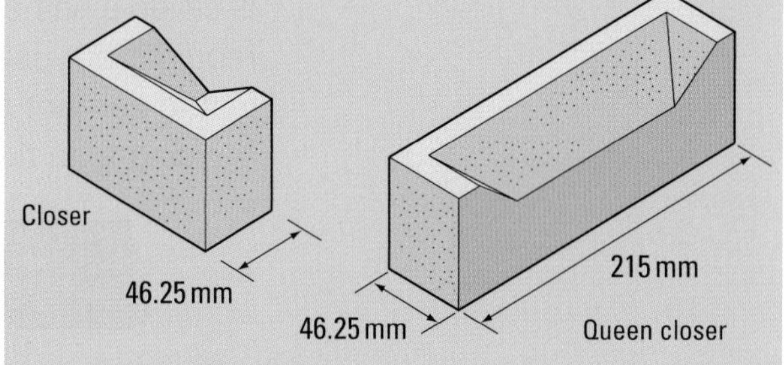 Closer 46.25 mm 46.25 mm 215 mm Queen closer Figure 7.15 Ordinary closer and queen closer
Batts	Various terms are used to describe parts of the brick. A **half batt** is a brick that has been cut in half across the width, with dimensions of 102.5 × 102.5 × 65 mm. A **three-quarter batt** has dimensions of 158.75 × 102.5 × 65 mm.	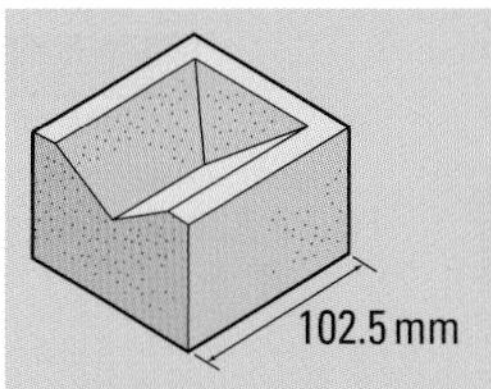 102.5 mm 158.75 mm Figure 7.16 Half batt 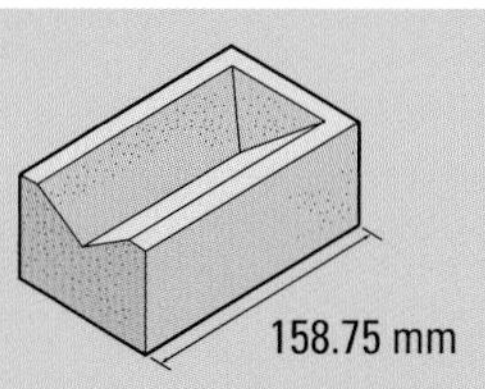 Figure 7.17 Three-quarter batt
Bed joints and courses	**Bed joints** are the horizontal joints of mortar between each course of brick. A **course** is the term that is used to describe a row, usually horizontal, of bricks.	
Cross, transverse and collar joints	A **cross joint** is the vertical joint between two bricks. A cross joint is also known as a **perpend** or perp. The **transverse joint** is also known as a **sectional joint**. These are at right angles to the face of the wall. **Wall joints** are those that run parallel to the face of the wall.	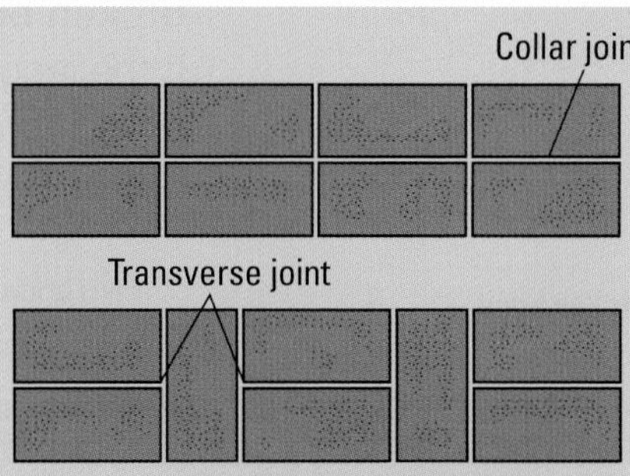 Collar joint Transverse joint Figure 7.18 Wall and transverse joints
T-junctions and cross junctions	A **T-junction** is where two walls meet, with a T-shaped formation. A **cross junction** is where two walls cross over one another. Where a wall finishes and has a wall face it is known as a **stopped end**.	 Figure 7.19 T-junctions  Figure 7.20 Cross junctions
Quoins and return angles	On the face side of a wall corner or angle, depending on which direction you are looking at the wall, the bricks that form the angle are known as **quoin headers** or **quoin stretchers**. At return corners the bricks forming angles of 90° are known as **square quoins**. Those with angles that are not 90° are known as **squint quoins**.	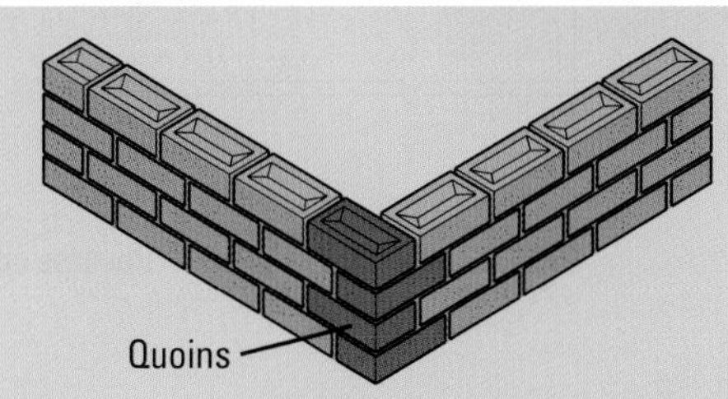 Quoins Figure 7.21 Return angle

Table 7.3 Terms used in brickwork

The simplest type of bonded wall is one where the bricks are all laid down as stretchers and they overlap one another by half a length (this is called a half-brick wall). This would be fine if only a half-brick wall was required but usually a one-brick wall is needed. It would not be suitable just to cement two half-brick walls together. They require better bonding than this to make them stable.

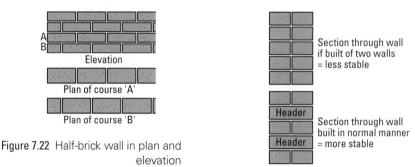

A
B
Elevation

Plan of course 'A'

Plan of course 'B'

Figure 7.22 Half-brick wall in plan and elevation

Section through wall if built of two walls = less stable

Header

Header

Section through wall built in normal manner = more stable

Figure 7.23 Two half-brick walls side by side and a bonded wall

There needs to be some kind of bonding across the two walls. This ensures that the load is distributed along the thickness and the length of the wall.

Broken bond

If the length of course of brickwork is not precisely the multiple of several brick lengths, then batts should be inserted. This is known as broken bond.

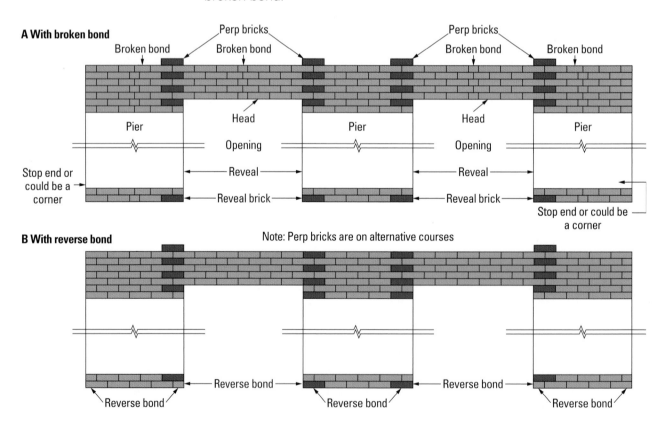

Figure 7.24 Marking out the bond

Dealing with broken bonds requires some planning. The position of the batt is not really important below ground level, but once ground level has been reached it is important to:

* accurately mark the position of any opening

* place the broken bond at the centre of any pier or opening to ensure that all the vertical joints will be plumb and it will also look better.

Half-brick bond (stretcher bond)

The bonding used for half-brick walls is where all the bricks are laid down as stretchers. Each brick overlaps the one directly below it by half of its length. This is obtained by starting each alternate course of bricks with a half-brick. This is probably the most common type of bond. The elevation and plan of the half-brick wall bonding can be seen in Figure 7.22.

English bond

An English bond is the strongest bond used in bricklaying as there are no straight joints in the wall. It is a quarter bond used in walls that are at least one brick thick.

English bond consists of alternate courses of headers and stretchers, with a queen closer next to the corner, producing a quarter bond. Some people think that English bond looks a bit boring and monotonous but it can be enhanced by using contrasting headers.

Flemish bond

Flemish bond is a type of quarter bond used for one-brick walls or over. It consists of alternate stretchers and headers in the same course. The header should be placed in the centre of the stretcher on the course below and above it. This creates a more attractive bond than English bond.

The bond is achieved by introducing a queen closer next to the corner header, giving you a quarter bond.

English bond

Elevation

Plan of course 2

Plan of course 1

Flemish bond

Elevation

Figure 7.25 English and Flemish bond

Maintaining industrial standards and working at height

Working to industrial standards means following clearly accepted guidelines as to the quality of work. These would include:

* erecting walls that are plumb levelled and finished

* ensuring that the wall is to gauge

* ensuring that the joints on all of the courses are plumb and true

* cleaning up the walling of splashes and excess mortar when finished.

When working at height

As the height of the brick wall increases it is impossible to carry out the work at ground level. This means that control measures need to be used to deal with any potential hazards. Working platforms are far safer than using ladders. Any working platform should have toe-boards and guard rails.

There is more information about working at height in Chapter 1.

Carrying out checks

There are a number of basic things that you should do in your preparation, actual construction and after construction:

* When you are working in cold weather never dampen the bricks before you lay them. This could cause the bricks to move around slightly as they dry out and the weather gets warmer.

* If you are constructing a wall that will need three or more packs of bricks you should select some bricks from each pack. This will reduce the risk of the wall having different bands of colour. There is no guarantee that the packs of bricks delivered all come from the same batch.

* Always use the same technique for cutting your bricks. On many sites there is a central cutting station. This ensures that all of the brick cuts are even and standard.

* It may not be possible to build the entire piece of walling over the course of a day or a few days. Poor weather might mean that you will need to leave it. To prevent damage to the wall and the joints you should always cover the work with a weatherproof sheet.

* You should make regular checks to see whether the length, height and finish of the wall looks accurate and similar.

* Slight differences in length or height can be easily dealt with. If the wall is only a matter of millimetres out this is not too great a concern.

* If there are any particular problems with the completed work then you should contact the site supervisor and report it. It will be the responsibility of others to speak to whoever needs to be consulted to resolve a problem. But you should not carry out any additional work until the problem has been resolved.

Setting out straight brick walls and corners to line

When basic structures are being set out you will see boards and lines fixed to pieces of timber that are embedded in the ground. These boards and lines are known as profiles.

The idea of the profiles is to establish corners and to act as a guide to show where the foundations need to be laid.

Once the profiles are in place the next job is to mark out the wall lines with chalk. This is achieved by holding or fixing one end of the line and then extending it to the finish position. The line is then snapped to leave a chalk outline on the surface (see Fig 7.27). This process is continued until all of the sides have been marked out. The process can be seen in Fig 7.26, Fig 7.27 and Fig 7.28.

Figure 7.26 Profiles and lines on site

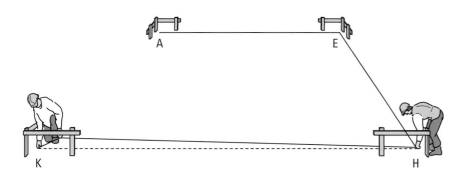

Figure 7.27 Marking outside AE

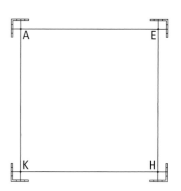

Figure 7.28 All sides marked out

CONSTRUCTING STRAIGHT BRICK WALLING AND RETURN CORNERS

This final part of the chapter looks at the ways in which straight brick walling and return corners are built. In each case you should begin as follows:

1. You need to identify and state exactly how the straight brick walling and return corner will be made. This means stating whether it is half-brick stretcher bond, one-brick walling or some other form of construction.

2. You then need to identify and state what type of bond is being used. Each of the practical tasks uses different bonds and over time these should become easily recognisable to you.

REED TIP

There's no use carrying your hard hat around – put it on your head!

3. You should always state why you need to carry out checks on the materials that you will be using, your tools and equipment and why it is necessary to make sure that your PPE is not only available but also in good condition.

Constructing straight brick walling and return corners in half-brick stretcher bond

This is the first set of practical tasks. They will show you how to produce the correct joint finishes, which include flushed and half-round. It is important to remember that you need to continually check the progress of the construction to make sure that it is in line with any instructions that you have been given.

Dry bonding is the term used for laying each brick in the first course without using any mortar joints either for the bed joint or the cross joint. Bricklayers need to do this in order to check that each course will work to exact brick sizes with no opening up or tightening up the cross joints required. It is also a way of checking whether cut bricks will be needed to complete each course. It is also good practice to do this if you have any window or door openings to consider higher up. This enables you to set out the bonding with these in mind.

KEY TERMS

Dry bonding

– laying bricks without applying a bed or cross joint in order to establish the bond and size of cross joints required for a given dimension.

PRACTICAL TASK

1. USE A LEVEL AND GAUGE LATH TO PLUMB AND GAUGE WALLS

OBJECTIVE

To get used to using a level for plumbing walls and a gauge lath for gauging each course of bricks. It will also help you recognise the basic terminology used in brickwork.

PPE

Ensure you select PPE appropriate to the job and site where you are working. Refer to the PPE section of Chapter 1.

TOOLS AND EQUIPMENT

Walling trowel

Pointing trowel

Spirit level

Lump hammer and bolster chisel

Builder's square

Jointing iron

Gauge rod or tape measure

STEP 1 Set out the mortar board in the given area.

PRACTICAL TIP

Some bricklayers say that 'the distance of your own two feet' is enough space to work in.

STEP 2 Dry bond the first course and mark both ends with a chalk.

Figure 7.29 Marking with chalk

PRACTICAL TIP

When dry bonding, a 10 mm spacer can be used between the bricks to give you the gap between the bricks for the cross joint.

STEP 3 Pick up and bed (lay with mortar) one brick against the positioned mark.

Figure 7.30 Pick up and bed one brick

STEP 4 Using a gauge rod, or if preferred a tape measure, press the brick down to the gauge mark or 75 mm.

Figure 7.31 Positioning the first brick

PRACTICAL TIP

It is best to level and plumb the brick after you have got it to gauge. It is pointless to level the brick and then check it for gauge – if it is too high then you will have to tap it down or if it is too low then you will have to take it up and put more mortar under it.

STEP 5 Check the brick for level and plumb and adjust if required but be careful not to knock the brick below the gauge.

Figure 7.32 Checking for level

STEP 6 Lay the second brick. Make sure you level across from the first brick to the second brick and do the same with the third brick.

Figure 7.33 Laying the second and third brick

STEP 7 Repeat the previous steps on each course until you complete the wall.

Figure 7.34 The second course

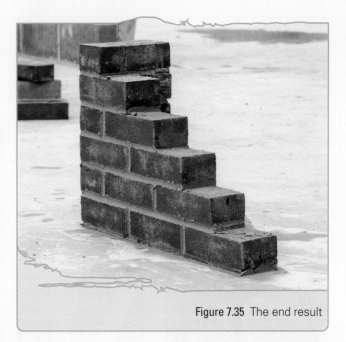

Figure 7.35 The end result

PRACTICAL TASK

2. BUILD A HALF-BRICK WALL BETWEEN PROFILES

OBJECTIVE

To practise laying bricks to a line.

The use of profiles means that this skill can be practised in isolation without being distracted by the additional work of building corners with a level.

Contrasting bricks can be added if you decide on a pattern beforehand and work it into the model as work progresses.

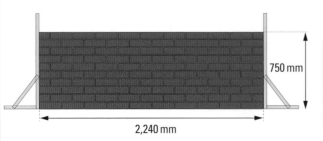

750 mm

2,240 mm

Figure 7.36 You will build a wall like this one

TOOLS AND EQUIPMENT

Brick trowel	Jointing iron
Lump hammer	Spirit level
Bolster chisel	Straight edge
Block/pins and line	Steel tape measure

PPE

Ensure you select PPE appropriate to the job and site where you are working. Refer to the PPE section of Chapter 1.

STEP 1 Set up the profiles so that they are plumb in both directions and 2,240 mm apart so that the wall is 10 bricks long.

Figure 7.37 Setting up the profiles

STEP 2 Dry bond the remaining eight bricks on the first course and lay to the line once they have been evenly spaced.

Figure 7.38 Dry bonding the wall

Figure 7.39 Laying to the line

PRACTICAL TIP

If you start at the lower end you will have to start cutting bricks lengthways to get them down to the line. If you start at the highest point you can always use extra mortar to bed bricks up at the longer end.

PRACTICAL TIP

Practise forming the perp joint at either end of the brick as this will help you to be able to work in both directions.

PRACTICAL TIP

When laying the beds, ensure that the mortar is a consistent width and height. The front of the bed should be flush and the back of the bed angled so that excess mortar does not squeeze out and drop to the back of the wall.

PRACTICAL TIP

If the blocks are correctly placed, they will be level with the line as you look at the course horizontally, and set back 1–2 mm from the line as you look from above. Make sure the bricks do not touch the line.

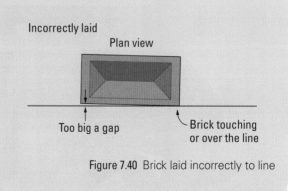

Figure 7.40 Brick laid incorrectly to line

STEP 3 If the floor is uneven, lay the first brick at the highest end and transfer the level to the other end using bedded bricks and a spirit level.

Figure 7.41 Levelling for high end

Figure 7.42 Transferring the levels – with two bricks

Figure 7.43 Transferring the levels – with all bricks

STEP 4 Mark the heights of the courses up the profile with either a tape measure or a gauge lath.

Figure 7.44 Marking the heights of the courses

STEP 5 Cut 2 half-bricks with a batt gauge, ensuring that the face is a crisp 90° cut and lay one at either end with the cut face turned into the wall. Then run in the remaining bricks.

Figure 7.45 Laying the ends

Cut away the excess mortar on the face with the inside edge of your trowel and pushing it forwards. This not only keeps the mortar from marking the face, but also helps to keep alignment along the bottom arris of the brick.

After the second course, you will be able to check if the bricks are 'creeping' forwards or backwards from their correct position by looking down the face of the brickwork and aligning them by eye with previous courses.

STEP 6 Repeat until the full height of the wall has been reached (750 mm).

Figure 7.46 Building up the wall (1)

If the jointing iron is leaving black marks this would indicate that the joints have been left too long before finishing.

STEP 7 Finish the wall with a half-round joint. Start by jointing all the vertical joints or perps and then running along the horizontal beds. Any holes should be filled and the horizontal beds should be unbroken by the perp joints.

Figure 7.47 Building up the wall (2)

Figure 7.49 Cutting away excess mortar

Figure 7.48 Building up the wall (3)

Figure 7.50 Pointing

Figure 7.51 A half-round joint

STEP 8 Brush the wall when it is dry enough not to leave brush marks. Brushing diagonally will help to reduce this.

Figure 7.52 Brushing the wall

Figure 7.53 Tipping in

Figure 7.54 The finished wall

PRACTICAL TASK

3. BUILD A HALF-BRICK RETURN CORNER

OBJECTIVE

To learn how to build a half-brick wall corner by mastering the basics of level, plumb and gauge.

Along with laying bricks to a line; these are some of the principal and fundamental aspects of laying bricks. Only when these elements have been achieved should you continue onto more sophisticated models.

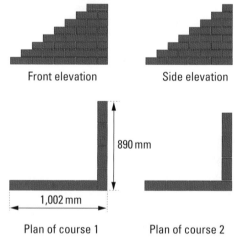

Front elevation   Side elevation

890 mm

1,002 mm

Plan of course 1   Plan of course 2

Figure 7.55 You will build a wall like this one

TOOLS AND EQUIPMENT

Brick trowel	Builder's square
Lump hammer	Jointing iron
Bolster chisel	Steel tape measure
Spirit level	

PPE

Ensure you select PPE appropriate to the job and site where you are working. Refer to the PPE section of Chapter 1.

STEP 1 Screed a thin layer of cement on the floor and mark a 90° angle with the blade of a brick trowel.

Figure 7.56 Marking a square in cement

PRACTICAL TIP

Indicate the corner with a line, in case any markings get obscured when the bed is laid.

Figure 7.57 Marking a line

Figure 7.58 Putting mortar down

STEP 2 Along one side, lay the first corner or 'quoin' brick and check for level and gauge.

Figure 7.59 Laying the first quoin brick

PRACTICAL TIP

It helps to check the gauge while the level is on the brick. This means that it can be altered to gauge while ensuring the brick is level at all times and it won't be moved out of gauge.

STEP 3 Lay the next three bricks to line until you have a line of four bricks that is 890 mm in length.

Figure 7.60 Laying bricks to line

Figure 7.61 Checking alignment

STEP 4 Lay four more bricks at 90° to the first four, making sure that they are level in relation to them.

Figure 7.62 Laying bricks at 90°

STEP 5 Check that the second side is 90° to the first side with a builder's square and move them accordingly if required.

Figure 7.63 Checking for square

STEP 6 Lay the quoin brick and set it for level, gauge and plumb in that order.

Figure 7.64 Laying the next quoin brick

PRACTICAL TIP

When plumbing, always check the header face first and then the side immediately next to it, before finally plumbing or aligning the far end of the brick.

The next stage can be done in two ways, so follow either Step 7 or Step 8.

STEP 7 Either the bricks can be laid in the same order as the first course, using a half-brick as a spacer in order to maintain bond.

Figure 7.65 Laying the second course

Figure 7.66 Checking the level of the second course

STEP 8 Alternatively place a half-brick at one end as a spacer and then lay the brick next to it, levelled and plumbed from the quoin brick. Then the remaining bricks can be laid in between, using the level as a straight edge to check alignment.

Figure 7.67 Using a half-brick as a spacer

STEP 9 Whether you followed Step 7 or Step 8, repeat this procedure until you reach the required height of eight courses.

Figure 7.68 Building up to eight courses

STEP 10 Point your corner with a half-round joint and when the mortar joints are sufficiently dry, use a soft brush to finish.

Figure 7.69 The finished corner

PRACTICAL TASK

4. BUILD A STRAIGHT HALF-BRICK WALL WITH TWO CORNERS

OBJECTIVE

To practise building corners and laying to the line.

Once this has been mastered, this skill is the basis for the majority of work on all following models. In addition, this model also allows you to practise some basic setting out which can be expanded on later.

This can be an extension of the previous model, as shown in the picture at Step 1.

TOOLS AND EQUIPMENT

Brick trowel	Builder's square
Lump hammer	Jointing iron
Bolster chisel	Steel tape measure
Spirit level	

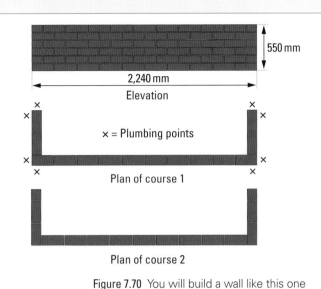

550 mm

2,240 mm

Elevation

× = Plumbing points

Plan of course 1

Plan of course 2

Figure 7.70 You will build a wall like this one

PPE

Ensure you select PPE appropriate to the job and site where you are working. Refer to the PPE section of Chapter 1.

STEP 1 Mark a line and lay 10 bricks out to a 2,240 mm length.

Figure 7.71 Extending the wall from the previous built corner

STEP 2 Dry bond these 10 bricks or six bricks if you are extending the previous model. Lay the first brick at the highest end of the floor and then transfer the level from one end to the other with either a spirit level or straight edge.

Figure 7.72 Transferring the level

PRACTICAL TIP

Remember to rotate the level 180° in order to eliminate any inaccuracies in the level.

STEP 3 Check the alignment with the straight edge and then re-check that length, level and alignment are all correct before walling commences.

Figure 7.73 Checking level end to end of second course

Figure 7.74 Checking plumb of return

STEP 4 Place two or three bricks on top of the two end bricks to stop them from moving when the lines are fitted with the blocks.

Figure 7.75 Checking end brick for plumb

STEP 5 The bricks can then be laid in to complete the first course. This should provide an accurate starting point from which all other parts of the model can be established. Proceed to build second return.

Figure 7.76 Laying in the first course

STEP 6 From this point, the return can be laid out level to the front of the wall and checked for 90° with a builder's square.

STEP 7 Then build up the corners to a height of 6 courses, one side at a time, using the techniques described in the previous practical task.

PRACTICAL TIP

Build the stop end as the corner is being built. This is far easier than racking the corner back and then adding the remaining bricks once the height of the corner has been reached. (Racking back is when you set each course back half a brick/block.)

STEP 8 Run the bricks in to the main body of the wall.

Figure 7.77 Running in the main wall

STEP 9 Point with a half-round joint and finish with a soft brush.

Figure 7.78 The finished wall

Constructing one-brick walling

This second batch of practical tasks focuses on building straight walls in one-brick walling. They will also look at the ways in which you should produce joint finishes.

PRACTICAL TASK

5. COMPLETE A SECTION OF ENGLISH BOND STRAIGHT WALL

OBJECTIVE

To practise constructing an English bond wall.

The model is to be built in local facing bricks.

Joint finish

Front = weather struck

Rear = flush from the trowel

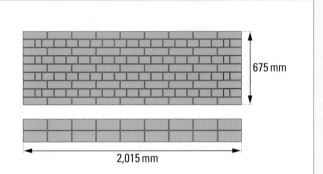

Figure 7.79 You will build an English bond wall like this one

675 mm

2,015 mm

TOOLS AND EQUIPMENT

Walling trowel	Jointing iron
Pointing trowel	Straight edge
Spirit level	Corner blocks and line
Lump hammer and bolster chisel	Bat gauge

PPE

Ensure you select PPE appropriate to the job and site where you are working. Refer to the PPE section of Chapter 1.

PRACTICAL TIP

Attention to the following points helps to maintain a good standard of work throughout all these practical exercises:

- Cut queen closers neatly and keep them regular in size.
- Remove mortar from the back of the bricks against the collar joint as it can prevent the backing up from being laid level. Keep perpends uniform and plumb because if the cross joints become too big then you will soon lose the bond, especially on the header course, and it will lead to straight joints on the face of the wall.
- When laying the back of the wall, avoid using too much mortar near the collar joint as it could prevent the bricks from being laid in line. When you tap the bricks back in place it may cause the face bricks to move out in front of the line.

STEP 1 Work out the amount of materials that you will need to complete the model.

Example

There are 120 bricks per square metre in a one-brick wall. To find out how many bricks you will need, you have to find out how many square metres are in the wall:

Length of wall (A) × Height of wall (B)

A × B = square metre (C)

You now have to multiply the square metres by 120:

C × 120 = N (number of bricks)

Don't forget that you might break or damage some of the bricks so you need to add some extra for wastage – normally 5%:

W (wastage) = N + 5% = Total number of bricks required

STEP 2 Cut eight closers using the appropriate tools. It is important to cut the closers to the correct size. If they are too big then it will throw out the bond on the header course.

PRACTICAL TIP

Queen closers are cut down the length of a brick. It is important to cut the closers to the correct size. If they are too big then it will throw out the bond on the header course and subsequently you will lose the bond and end up with straight joints. Do not just cut the brick down the middle as this will make the closer too wide so it will be difficult to keep the bond correct.

STEP 3 Set out and dry bond the first course to the correct length. Check the length of the wall.

STEP 4 Lay the first brick at one end to gauge and level. Now lay the second end brick.

Using a straight edge and level, level the second end brick with the first.

PRACTICAL TIP

When working in foundations never assume that the foundation concrete is level. Always check it before you start laying bricks and always start at the highest point. It is easier to bed the bricks up with more mortar than trying to cut the bricks down lengthways.

STEP 5 Using corner blocks and a line lay the rest of the course on the face of the wall.

PRACTICAL TIP

At this point the two corner bricks will still be 'wet', i.e. not set, so you will have to set some bricks on top of the corner bricks so that they will not move.

PRACTICAL TIP

When laying the bed joint for the back of the wall, always push any excess mortar away from the bricks already laid because at this point you need to check that it is one brick wide. You do this by laying a brick across the wall of the bricks that have been previously laid. If there is too much mortar between the two bricks (collar joint) then you will push out the face brick as you try to tap the rear brick to the correct width.

Figure 7.80 Weighing down the corner bricks

Figure 7.81 Removing excess mortar from the back of the wall

Figure 7.82 Checking the width of the wall

STEP 7 Wall the bricks to a line between the courses. Joint the wall as the work proceeds

STEP 8 Repeat Steps 4 and 5 until the wall is the correct height.

STEP 6 Build two small corners (five courses) at each end. Joint the corners as the work proceeds. Make sure that you keep the bond otherwise when you run the main wall you will not fit all the bricks in. Make sure that you place the header over the centre of the joint on the stretcher course below.

PRACTICAL TASK

6. COMPLETE A SECTION OF FLEMISH BOND STRAIGHT WALL

OBJECTIVE

To practise constructing a Flemish bond wall.

The model is to be built in local facing bricks.

Joint finish:

Front = half round

Rear = flush from the trowel

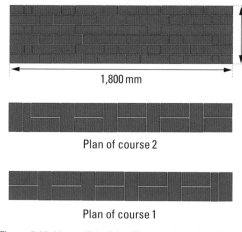

450 mm

1,800 mm

Plan of course 2

Plan of course 1

Figure 7.83 You will build a Flemish bond wall like this one

TOOLS AND EQUIPMENT

Walling trowel	Builder's square
Pointing trowel	Jointing iron
Spirit level	Corner blocks and line
Lump hammer and bolster chisel	Bat gauge

PPE

Ensure you select PPE appropriate to the job and site where you are working. Refer to the PPE section of Chapter 1.

STEP 1 Work out the amount of materials that you would need to complete the model.

Example

There are 120 bricks per square metre in a one-brick wall. To find out how many bricks you will need, you have to find out how many square metres are in the wall:

Length of wall (A) × Height of wall (B)

A × B = square metre (C)

You now have to multiply the square metres by 120:

C × 120 = N (number of bricks)

Don't forget that you might break or damage some of the bricks so you need to add some extra for wastage – normally 5%:

W (wastage) = N + 5% = Total number of bricks required

STEP 2 Cut the correct number of queen closers to the correct size using the appropriate tools.

STEP 3 Set out and dry bond the first course.

STEP 4 Lay the first brick at one end to gauge and level. Now lay the second end brick.

Using a straight edge and level, level the second end brick with the first.

STEP 5 Using corner blocks and line, run in the first course.

STEP 6 Build two small corners (five courses) at each end. Make sure that you keep the bond.

STEP 7 Using corner blocks and line, run in the courses between the corners.

STEP 8 Repeat Steps 6 and 7 until you have reached the correct height.

Figure 7.84 Close up of the detail of the brickwork

PRACTICAL TASK

7. COMPLETE AN ENGLISH BOND RETURN CORNER

OBJECTIVE

To practise building an English bond return corner.

The model is to be built in local facing bricks.

Joint finish

Front = weather struck

Rear = flush from the trowel

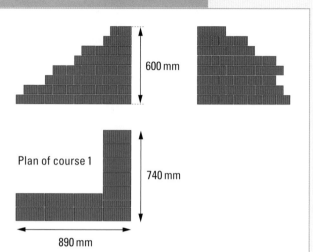

Plan of course 1

600 mm

740 mm

890 mm

Figure 7.85 You will build an English bond return corner like this one

TOOLS AND EQUIPMENT

Walling trowel	Builder's square
Pointing trowel	Jointing iron
Spirit level	Corner blocks and line
Lump hammer and bolster chisel	Bat gauge

PPE

Ensure you select PPE appropriate to the job and site where you are working. Refer to the PPE section of Chapter 1.

STEP 1 Work out the amount of materials that you would need to complete the model. For this model you need to take the average number of bricks half-way up the corner, in this case, course four as the basis for your calculation.

Example:

N (Number of brick required) = Number of bricks per course = A

B × Number of courses

Don't forget that you might break or damage some of the bricks so you need to add some extra for wastage – normally 5%

W (wastage) = N × 5%

Total number of bricks required = N + W

STEP 2 Cut seven closers to the correct size using the appropriate tools.

PRACTICAL TIP

Always lay the queen closers with the full face of the bed of the brick down. If you lay the closers 'frog' down you only have three edges and the closer will tend to tilt and will show up in the wall.

Figure 7.86 Queen closer bedded wrong way up

Figure 7.87 Queen closer bedded right way up

STEP 6 Now measure between the two marks. It should be 1,500 mm if the corner is 90°.

If it does not measure 1,500 mm then the second leg needs to be adjusted as you cannot adjust the face side.

STEP 7 Set out and dry bond the first course to the correct length as shown on the drawing. Check the length of the wall.

STEP 3 Using a builder's square set out a 90° angle and check it using the 3:4:5 method for accuracy.

STEP 8 Lay the first course and check it for square before continuing.

PRACTICAL TIP

It is important at this stage to check the accuracy of the corner. This can be done by using the 3:4:5 method. Take three straight lines: one 300 mm long, one 400 mm long and the third 500 mm long and join them together to make a triangle. The angle opposite the longest side will be 90°.

You can extend all sides in preparation, and the longer the measurement used, the more accurate the corner will be. But you must use the same unit to multiply each side e.g.:

Multiply 300 mm × 3 = 900 mm
400 mm × 3 = 1,200 mm
500 mm × 3 = 1,500 mm

PRACTICAL TIP

At this point it is best to check that you haven't made the joints between the headers too big. Dry bond the stretchers along the top of the header course and make sure that the joint is in the centre of the header below. Make any adjustments at this point.

STEP 9 Complete the rest of the corner and joint the wall as the work proceeds. Make sure that any toothing is kept clean and free from debris and that none of the bricks are tilting down.

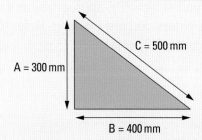

Figure 7.88 The 3:4:5 method

Figure 7.89 The finished wall

STEP 4 Now accurately measure 900 mm along the face of the wall and mark it on the floor. Recheck the measurement.

STEP 5 Measure along the second leg of the corner 1,200 mm and mark it on the floor as before.

PRACTICAL TIP

Toothing is a method of extending the wall at a later date. Bricks are left out on alternate courses and the extension can be toothed in later. Many architects do not like this method of extending walls as you cannot always guarantee that the resulting joints will be completely full, leaving a weak point in the wall.

PRACTICAL TASK

8. BUILD A FLEMISH BOND RETURN CORNER

OBJECTIVE

To complete a Flemish bond return corner to the required standards.

The model is to be built in local facing bricks.

Joint finish

Front = half-round

Rear = flush from the trowel

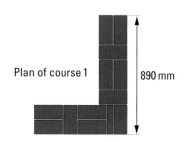

Side elevation

450 mm

Plan of course 1

890 mm

Figure 7.90 You will build a Flemish bond return corner like this one

TOOLS AND EQUIPMENT

Walling trowel	Builder's square
Pointing trowel	Jointing iron
Spirit level	Bat gauge
Lump hammer and bolster chisel	

PPE

Ensure you select PPE appropriate to the job and site where you are working. Refer to the PPE section of Chapter 1.

STEP 1 Work out the amount of materials that you would need to complete the model. See Step 1 on page 202 (English bond return corner) for guidance on how to do this.

STEP 3 Using a builder's square set out a 90° angle and check it using the 3:4:5 method for accuracy. See page 203 to remind yourself how to do this.

STEP 2 Cut the queen closers to the correct size. See Step 2 on page 198 (English bond straight wall) for guidance on how to do this.

STEP 4 Lay the corner stretcher brick to gauge and level both ways. Lay the remaining bricks along the face, levelling them with the corner brick.

STEP 5 Lay the return course, starting with a queen closer next to the corner brick. Check the corner with the builder's square for accuracy.

PRACTICAL TIP

At this point it is best to check that you have not made the joints between the bricks too big. Dry bond the second course along the top of the first course and make sure that the header is in the centre of the stretcher below. Make any adjustments at this point.

STEP 6 Complete the rest of the corner. Joint the wall as the work proceeds

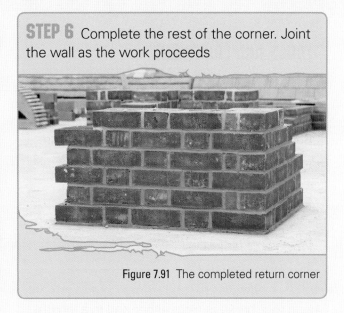

Figure 7.91 The completed return corner

Forming junctions in brick walls

In many cases it is necessary for you to create junctions in brick walls. These can be junctions that attach brick walls to existing structures, or might be a simple T-shaped junction. Some are more complex and require you to make piers. These practical tasks show you how to form the necessary ties. The correct joint finishes are also covered, as are ways in which you can check your work to see if it meets the instructions that you have been given.

PRACTICAL TASK

9. BUILD A HALF-BRICK WALL WITH A T-JUNCTION

OBJECTIVE

To practise connecting two half-brick walls without losing bond or creating any straight joints that will affect the strength of the wall.

This depends on the use of the three-quarter batt, which is the only cut brick that should vary from a set dimension, and allows it to be adapted so that the correct bond can be maintained.

Contrasting coloured bricks may be used to define detail areas such as the three-quarter bricks.

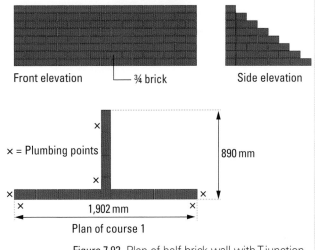

Front elevation   ¾ brick   Side elevation

x = Plumbing points

890 mm

1,902 mm

Plan of course 1

Figure 7.92 Plan of half-brick wall with T-junction

TOOLS AND EQUIPMENT

Brick trowel	Straight edge
Lump hammer	Jointing iron
Bolster chisel	Steel tape measure
Spirit level	Builder's square
Block/pins and line	

PPE

Ensure you select PPE appropriate to the job and site where you are working. Refer to the PPE section of Chapter 1.

STEP 1 Mark a line 1,902 mm long and dry bond four bricks in from either end to a length of 890 mm. This should leave the correct spacing for the junction to be attached.

STEP 2 Lay the brick at the highest end (if there is one) and transfer the level to the other end brick.

STEP 3 Weight the end bricks, attach the line and run in the stretcher bricks so that there is a 120 mm opening in the middle that will enable the junction wall to be placed.

Figure 7.93 Dry bonded bricks with a 120 mm opening

STEP 4 Lay the four bricks in the junction to 890 mm from the face of the wall and allow for 10 mm joints on either side of the connecting brick.

STEP 5 Check the junction for 90° using a builder's square.

Figure 7.94 Laying the first and checking it's square

PRACTICAL TIP

Check only one side of the junction for square and keep this as the side to be plumbed. Check for square from the face of the main wall as it will not be as accurate from the internal corner.

STEP 6 Build the corners on the face wall up to height. Leave the three-quarter bricks on the second course until the wall is being run in on the face.

Figure 7.95 Building the corners on the face wall to height

STEP 7 Attach the lines and prepare to run in the face brickwork.

Figure 7.96 Attaching the lines

PRACTICAL TIP

The three-quarter bricks can be marked by marking a centre line down the middle of the stretcher that connects the junction. Place a 5 mm mark on either side. Then place a 10 mm mark from the end bricks of the corners. The distance between will be the correct size of the three-quarter bricks. This will also ensure that the joint is correctly placed in the centre with the header faces showing on the face side of the wall.

Figure 7.97 Marking the three quarter cuts

STEP 8 The bricks in the junction can be laid out with the level. Only one side should be plumbed and this should be the face that was checked for square.

Figure 7.98 Laying out the junction bricks

STEP 9 Continue running in and extending the junction with the level until the full height of the model is reached.

PRACTICAL TIP

Run all the bricks that are possible to the line first before laying the bricks that need to be laid to the spirit level.

STEP 10 Finish the model with a half-round joint and brush when ready.

Figure 7.99 The finished T-junction

10. BUILD A HALF-BRICK WALL WITH ATTACHED PIERS AND CONCRETE COPING

OBJECTIVE

To practise the skills of the earlier models, but with the addition of basic attached piers capped off with concrete copings.

The one brick pier is also built with a king closer and mitred batt which will eliminate the occurrence of any straight joints within the pier and wall.

PPE

Ensure you select PPE appropriate to the job and site where you are working. Refer to the PPE section of Chapter 1.

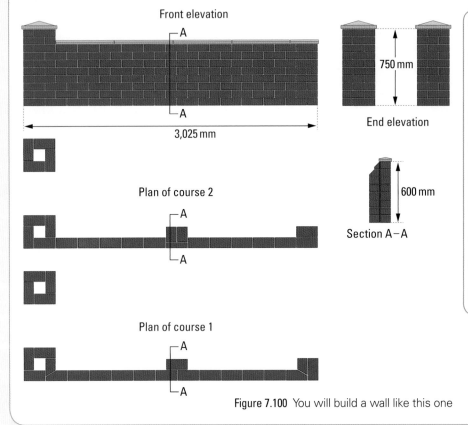

Figure 7.100 You will build a wall like this one

TOOLS AND EQUIPMENT

Brick trowel

Lump hammer

Bolster chisel

Spirit level

Block/pins and line

Straight edge

Jointing iron

Steel tape measure

Builder's square

STEP 1 Mark a line and dry bond 10 bricks 3,025 mm in length. Lay the first brick at the highest end of the floor and then transfer the level from one end to the other with either a spirit level or straight edge.

PRACTICAL TIP

Remember to rotate the level 180° in order to eliminate any possible inaccuracies in the level if using this method.

STEP 2 Check the alignment with the straight edge and then re-check that length, level and alignment are all correct before walling commences.

STEP 3 Place two or three bricks onto the top of the two end bricks to stop them from moving when attaching the corner blocks and line.

STEP 4 The bricks can then be laid in to complete the first course. This should provide an accurate starting point from which all other parts of the model can be established.

STEP 5 At the required end, lay a brick at 90° to form the start of the one-and-a-half brick pier. Level this brick to the main body of the wall and check for square and length along that side (327 mm).

STEP 6 Then lay the remaining bricks to form the pier and check for level, square and correct dimensions.

STEP 7 At the other end, lay a brick adjacent to the last brick, allowing for a 10 mm perp joint. Check for square and level to the main body of the wall.

STEP 8 Place the intermediate pier brick, again allowing for a 10 mm perp joint, checking for level and the correct placement to the main body of the wall.

STEP 9 Then the corners can be raised so that the wall can be run in. Starting at the one-and-a-half-brick end, the pier can then be built up as the wall is racked back where it runs into the main body of the wall.

STEP 10 Build the end with the one-brick pier with a king closer and mitred batt on alternate courses in order to eliminate any straight joints. Build the corner up to the height required, adding a toothed joint if necessary.

STEP 11 Run in the main body of the wall. Opposite to the intermediate pier, remember to add a queen closer. This will allow the three-quarter bricks to form the pier section on that course.

STEP 12 On the two courses before the wall is reached, add cant bricks. These have 45° chamfered faces between their stretcher and header surfaces. They reduce the width of the wall from the pier to the finished height where the copings will be added.

STEP 13 Lay the end coping and the coping closest to the brick-and-a-half-pier. Make sure that they are in line and that each side is equally spaced over the width of the wall. The remaining copings can then be run to the line.

STEP 14 Bed the pier capping on and ensure that it is squarely positioned with a tape measure. Check for level on the underside of the capping.

STEP 15 Point the wall with a half-round joint and brush it diagonally when the mortar is dry enough for brush marks to be avoided.

TEST YOURSELF

1. When you are about to begin bricklaying, which of the following is not a hazard to consider?

    a. The weight of the materials being used

    b. The walling can be unstable until the mortar sets

    c. You may have to be working at height

    d. The walling may not be to gauge

2. What is CAD?

    a. Computer Aided Design

    b. Construction Aided Design

    c. Computer Aided Demolition

    d. Construction And Drawings

3. What kind of machine should be used to lift packs of bricks onto scaffolding?

    a. Brick trolley

    b. Hod

    c. Telescopic handler

    d. Crane

4. In order to work out the area of a section of wall, which two measurements do you need?

    a. Length and height

    b. Length and width

    c. Width and height

    d. The two diagonals

5. Safety boots should have steel toecaps, but what other safety feature should they also have?

    a. They should not have laces

    b. They should have rubber soles

    c. They should have steel mid-soles

    d. They should be high-vis

6. What is toothing?

    a. A type of finishing tool

    b. When bricks are cut out at the end of a wall that is going to be extended

    c. When you lay each brick of the first course without using any mortar joints

    d. Another name for English bond

7. Which of the following types of brick are the most dense and are often used for damp-proof courses and retaining walls?

    a. Facing

    b. Common

    c. Clay

    d. Engineering

8. What is the depression at the top of a brick called?

    a. Bed

    b. Arris

    c. Frog

    d. Batt

9. Where would you find a return angle or return?

    a. At a junction between two walls

    b. At the top of a wall only

    c. In the first course of bricks only

    d. In the foundations only

10. In a stretcher bond, what distance is the overlap of the bricks?

    a. Quarter brick

    b. Half brick

    c. Whole brick

    d. Three-quarter brick

Chapter 8

Unit CSA–L1Occ21
CONSTRUCT CAVITY WALLING

LEARNING OUTCOMES

LO1/2: Know how to and be able to prepare resources for constructing cavity walling to the given instructions

LO3/4: Know how to and be able to set out to construct cavity walling

LO5/6: Know how to and be able to construct cavity walling and return corners to the given instructions

INTRODUCTION

The aims of this chapter are to:

* help you interpret and understand information

* help you to set out cavity walling to line

* help you to construct cavity walls.

PREPARING RESOURCES

A cavity wall consists of two walls called leaves. The leaves have a gap between them. This gap is known as the cavity.

The outer wall or leaf can be made of brick and the inner leaf, which is normally 100 mm wide, made of blockwork. The walls are anchored together using wall ties. This gives the two relatively thin walls greater strength.

The main point behind building a cavity wall is to prevent damp from penetrating into the building. The cavity provides a break between the wall exposed to the weather and the inner wall. At the same time the cavity can keep heat in the building during the cold months and prevent heat from entering the building during the summer.

Hazards, health and safety and risk assessment

Many of the potential hazards in building cavity walling are the same as those that apply to the building of other structures, such as brick walls or block walls. You should refer to Chapter 6 and Chapter 7, where potential hazards associated with building block walling and brick walling were discussed. You should also look at Chapter 1, which dealt with health and safety in general terms.

The other additional concern with regard to building cavity walling, particularly if it is being used in an entire dwelling rather than an extension, is working from height.

Information sources and drawings

Drawings are covered in detail in Chapter 2. The information about sources and drawings is basically the same as that covered in Chapter 6 and Chapter 7.

Confirming instructions

It is always important to check and confirm any instructions you are given.

You should always make sure that you fully understand what is expected of you.

There is more detail about checking that you understand your instructions in Chapter 7.

Safe methods of handling bricks and blocks

The advice given in Chapter 6 on blocks and in Chapter 7 on bricks obviously also applies here. It is important to follow the advice given in these chapters as you will be working with both kinds of material when you build cavity walling.

PPE

The Personal Protective Equipment at Work Regulations apply when building cavity walling. The advice given in Chapter 7, also relates to the building of cavity walling.

Resources

The resources needed to build cavity walling obviously include what was needed to build both block walling and brick walling. You should refer to Chapter 6 and Chapter 7.

Calculating resources

Chapter 6 contained detailed information about how to calculate the blocks needed to build block walling. Chapter 7 covered similar calculations required to work out how many bricks are needed. You should refer to both of these sections when calculating resources for cavity walling.

Cutting bricks and blocks by hand

The main thing to remember is that any bricks you are cutting are likely to be facing bricks, which will be visible.

The blockwork you are cutting is most likely to be covered by plasterboard or some other covering.

The ways in which you can prepare and cut bricks and blocks are both covered in Chapter 6, for blockwork and Chapter 7, for bricks.

Protecting the work and the surrounding area from damage

As we have seen newly built brick and blockwork can be vulnerable. You should follow the advice on the best ways to protect your work and prevent the surrounding area from being damaged.

KEY TERMS

Drip

– this is a bend or kink in the wall tie that prevents water from crossing the wall tie.

SETTING OUT TO CONSTRUCT CAVITY WALLING

The building of cavity walls needs careful and precise design and construction. This means remembering the following:

* The inner and outer leaves need to be tied together with wall ties.

* The wall ties need to either slope down towards the outer leaf or be level.

* The **drip** should be in the centre of the cavity and point downwards.

* The specification will state exactly which kind of wall ties are to be used. This could be important because particular wall ties might be needed for the insulation material that will be put into the cavity.

* There should be no mortar droppings inside the cavity. This is particularly important if insulation material is to be used.

* All joints on the face work need to be properly filled with mortar in order to stop water from penetrating.

* A damp-proof course membrane needs to be put in place and if there are any lintels cavity trays need to be installed. This also means that weep holes are necessary to drain the water from the cavity trays.

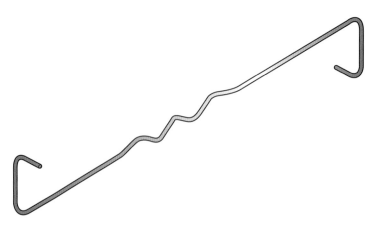

Figure 8.1 A general purpose wall tie

We look in more detail at damp-proof courses and wall ties used in cavity walling a little later.

Positioning bricks, blocks, mortar and components

The advice given in Chapter 6 and Chapter 7 applies to the building of cavity walling a little later in this chapter.

Generally it is important to have the components you need ready at hand, without stacking up too many components that could hinder the work.

Sequence of work and walling heights

In the past the outer leaf of brickwork was generally built first and the blockwork added afterwards. The more common method now is to build the blockwork first. This is mainly because the cavities are now filled with insulation. The cavity insulation is held in place using clips that are fixed to the wall ties on the inner leaf. You would also need scaffolding on the inside if you built the outer leaf first.

The general advice given is that the brick courses are gauged at 75 mm per course. If you follow this advice every third course of bricks will run level with a course of blockwork. This is ideal, as it will allow wall ties to be fitted in both the bed joints.

The advice given in Chapter 6 regarding blockwork and in Chapter 7 on brickwork about the number of courses that should be built at any time is also relevant for cavity walling.

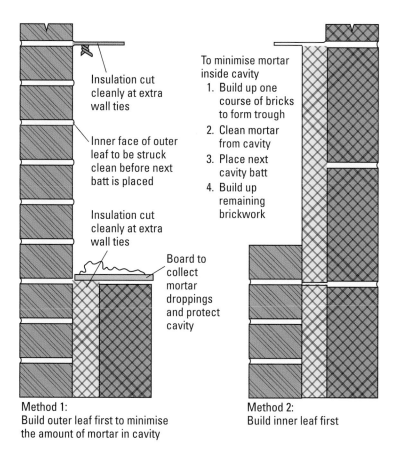

Insulation cut cleanly at extra wall ties

Inner face of outer leaf to be struck clean before next batt is placed

Insulation cut cleanly at extra wall ties

To minimise mortar inside cavity
1. Build up one course of bricks to form trough
2. Clean mortar from cavity
3. Place next cavity batt
4. Build up remaining brickwork

Board to collect mortar droppings and protect cavity

Method 1:
Build outer leaf first to minimise the amount of mortar in cavity

Method 2:
Build inner leaf first

Figure 8.2 A cavity filled with insulation

Industrial standards

Working to industrial standards means following clearly accepted guidelines and maintain a certain quality of work. These would include:

* erecting walls that are plumb levelled and finished

* ensuring that the wall is to gauge

* ensuring that the joints on all of the courses are plumb and true

* cleaning up the walling of splashes and excess cement or mortar when finished.

Safe working practices when working at height

As the height of the brick wall increases it is impossible to carry out the work at ground level. This means that control measures must be used to deal with any potential hazards. Working platforms are far safer than using ladders. Any working platform should have toe-boards and guard rails.

There is more information about working at height in Chapter 1.

Setting out straight brick and block walls to line

The advice given about setting out brick and blockwork to line has already been covered in Chapters 6 and 7. The additional consideration for cavity walling is to ensure that you maintain a regular space between the inner and outer leaves of the cavity wall. This is important because the wall ties need to be firmly fixed to both walls. In addition, the cavity is there to form a barrier and this should not be affected along the length of the wall.

CASE STUDY

South Tyneside Homes

South Tyneside Council's
Housing Company

Attention to detail

Gary Kirsop, Head of Property Services at South Tyneside Homes, says:

'The important thing is to make sure that your dimensions at the bottom are right. Setting out your work area properly will help make sure the wall is perfect, and then you'll be plumb as you're coming up.

You also need to keep the area clean so it's not full of excess mortar. You don't want to be spending extra time knocking bricks out at the bottom to rake out your cavities, so take your time to make sure it's clean all the way up.

Attention to detail and precision is key – make sure you're putting your cavity wall ties in the right position.

Make sure you've calculated how many bricks and blocks you need so you don't end up with an excess of materials around you. It's vital to keep your area clean and tidy, not just for health and safety, but also so you don't risk damaging materials because you've over-calculated. It's critical that you get the right sizes and dimensions – if you get this wrong and end up with too many materials, it is a costly mistake. These days the costs of materials is horrendous in construction, and if you've got too many, it's not just the cost of the supplies themselves, but the transportation to move them on, and some companies won't even take them back. Many suppliers will tell you it's too risky because they might already be chipped or damaged.'

CONSTRUCTING STRAIGHT CAVITY WALLING AND RETURN CORNERS

When you are constructing cavity walling it is very important to mark out the position of the walling. Even slight measurement mistakes can prove to be expensive. They are also very difficult to put right.

The walls need to be plumb and levelled. They need to be to gauge and, as the inner and outer leaves are tied together, the joints on each of the courses need to be level and true.

Constructing cavity walls means that you are often using both bricks and blocks. The two different walls may have different bonds and joint finishes. In addition to this, you also need to make sure that debris and mortar is not left inside the cavity wall.

Damp-proof barriers

Damp-proof barriers are a major way to stop moisture from entering a building. Moisture can come up from the ground, so a damp-proof course placed a minimum of 150 mm above ground level is vital to prevent this.

The cavity between the inner and outer leaves, along with other components such as **cavity trays**, prevents water from getting through the walls.

There is also the chance that moisture could come down from the top of the wall, particularly around openings. This is prevented by putting a a cavity tray above the lintel.

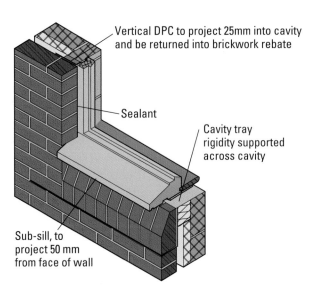

Figure 8.3 Using a cavity wall tray beneath an opening

Labels in figure:
- Vertical DPC to project 25mm into cavity and be returned into brickwork rebate
- Sealant
- Cavity tray rigidity supported across cavity
- Sub-sill, to project 50 mm from face of wall

The stretcher bond is the most common bond used for both the outer and inner leaves of cavity walling. Sometimes, when a cavity wall is being built as part of an extension, the outer leaf brickwork may have to use a different bond in order to match the existing building.

The bond is the particular pattern in which the bricks are laid in a wall. The idea with any bond is to ensure that the weight of the wall is evenly distributed along its whole length. This may mean setting out the bricks dry at first. This will show you how many full size bricks or blocks will be needed and whether you will need to cut any.

Usually the choice of bond is determined by the purpose of the wall and the strength that it needs.

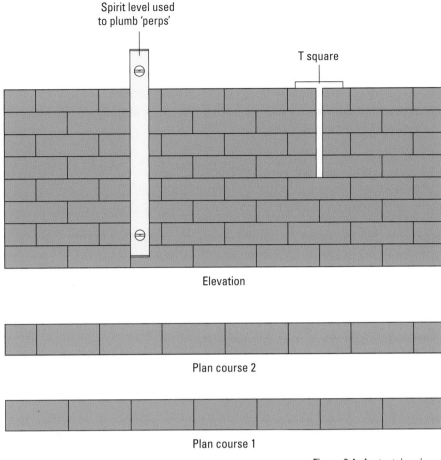

Elevation

Plan course 2

Plan course 1

Figure 8.4 A stretcher bond

Broken bonds

If you do need to use cut bricks then this is known as a broken bond wall (see Chapter 7). In most situations walls will be designed so that they work to match brick sizes. This is not always possible, however, and so cut bricks will be needed. Different sizes of brick will be needed to either open up or close the ideal overlap of the bricks on different courses.

Joint finishes

Jointing will depend on the desired look of the wall. The timing of jointing will depend on a number of factors:

* Moisture – if the bricks are wet then you will have to wait to joint until they have dried out.

* Temperature – on hot days joints will dry out quickly, as the bricks will absorb heat. On colder and wetter days joints may take time to dry out, as will the bricks.

* Time of day and progress of work – brick courses that you have already laid earlier in the day will be drier than those that you have worked on recently.

* Type of brick – the softer the brick, the more moisture it is likely to soak up, which in turn means that the joints will take longer to dry out.

PRACTICAL TIP

Try touching the joint to see how dry or wet it is. If the joint is too wet then the mortar will drag and the effect will be quite messy.

Pointing is the process of filling the mortar joints and there are several different ways in which the desired effects can be achieved after the wall is completed, which we look at in the next section. However, it is important that pointing is carried out slowly and carefully as this will be the finished look on the wall.

You should always remember the following things when pointing:

* Always start your pointing at the top of the wall. This is to ensure that any loose mortar does not fall onto freshly pointed areas.

* Ensure that the raking out has been done properly and that there is no loose mortar or dust in the joint.

* Remember to dampen the bricks around the area that you are pointing. You need to bear in mind that different bricks absorb water in different ways. Ensuring that the bricks are dampened will mean that your mortar will stick more easily.

* Once you have completed a section of pointing and it is dry, use a brush to clean off any dust.

PRACTICAL TIP

Various specialist pointing tools can be used. A pointing trowel can be used to make cuts in the mortar, or a tool known as a Frenchman is used to cut straight lines.

There are at least six different types of joint finish that you can achieve when pointing. These can be seen in the following table.

Joint finish	Description
Weather struck	The joint slopes downwards to encourage rainwater away from the joint and down the face of the brick. It is mainly used outdoors.
Ironed or tooled (half-round recessed)	One of the quickest jointing finishes – a shallow semi-circular indent is made in the joint.
Recessed	During pointing the mortar is compressed into the joint to a depth of around 4mm.
Flush	The mortar completely fills the joint up to the face of the brick.
Weather struck with cut pointing	The joint angles downwards and protrudes out beyond the face of the brickwork.
Reverse struck	This is the complete opposite of the weather struck joint. The joint angles inwards from the top. Used indoors only, as water will stand on the ledge.

Table 8.1 Different types of joint finish

Keeping wall cavities clean

Water penetration can be a problem in cavity walling. The dampness or moisture affecting the outer leaf can cross the cavity if there is what is known as a bridge. Bridges can be formed in three ways:

* mortar droppings inside the cavity

* poor placement of wall ties

* problems with the damp-proof course membrane.

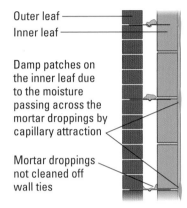

Figure 8.5 Dirty wall ties can cause damp patches in a wall

It is important to keep the wall cavities clean. When building the walls it is likely that mortar droppings will collect inside the wall cavity. They could:

* collect at the bottom of the cavity wall to a level that is above the damp-proof course

* collect on top of wall ties, which makes the tie drip ineffective.

The best way to avoid mortar from dropping into the cavity is to be a little more careful when you spread the mortar. You should try to cut back the inner face of the bedding joint to stop mortar from falling into the cavity.

Sometimes it is good practice to use a cavity batten. This is very simple and is made up of a length of timber that is held in place inside the cavity by two lengths of either wire or rope. The batten can be raised up and cleaned off when necessary.

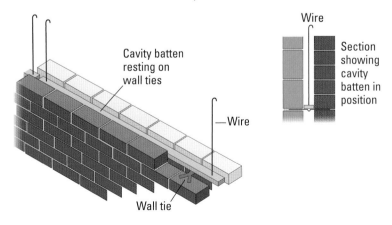

Figure 8.6 Method of keeping a cavity wall clean during construction

221

Cavity wall ties

Cavity wall ties are used to give strength to the walls. You will need some skill and common sense when you are fitting the wall ties. Building Regulations and the Masonry Code of Practice suggest that they should be:

* vertically spaced every block course at openings

* spaced at a distance of 900 mm maximum horizontally and 450 mm vertically.

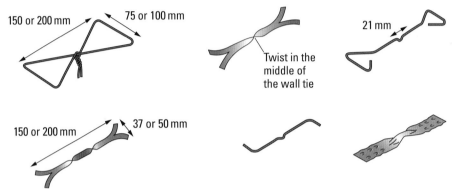

Figure 8.7 Cavity wall ties

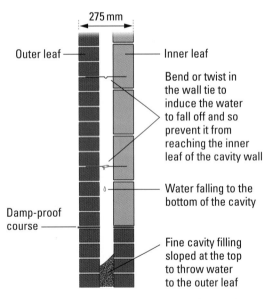

Figure 8.8 Ties trapping the passage of water from the outer and inner leaf

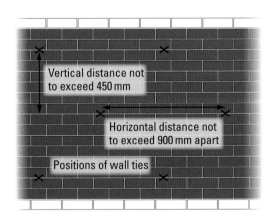

Figure 8.9 Elevation of a cavity wall showing the wall ties staggered throughout the wall

The wall ties need to be bedded in at least 50 mm to each wall leaf. This will ensure strength. The other considerations about placing the ties are:

* the drip needs to be pointing downward and at the centre of the cavity

* the ties need to be pressed down into the mortar bed firmly

* do not push the ties into joints as they will not tie the two leaves together properly

- if the ties are not horizontal then they should never incline towards the inner leaf but always the outer leaf

- never bend ties and always clean mortar droppings from them

- try to keep the gauge of the brickwork consistent and the thickness of the joints

- only ever use the specified type of wall tie.

REED TIP

Good communication is about listening – not just talking – and understanding the different ways people prefer to communicate.

Constructing straight cavity walling

The following practical tasks show you how to build different types of cavity walling. You should use both of these practical tasks to learn and practise the processes and skills required. These will help you to construct the cavity walling to given work instructions and produce the required joint finishes.

PRACTICAL TASK

1. BUILD A STRAIGHT CAVITY WALL

OBJECTIVE

To practise the skills required to build a basic cavity wall that is used in the majority of houses built in the United Kingdom today.

Once you have mastered the basic techniques, you can move on to the later models, which incorporate the installation of damp-proof courses and insulation.

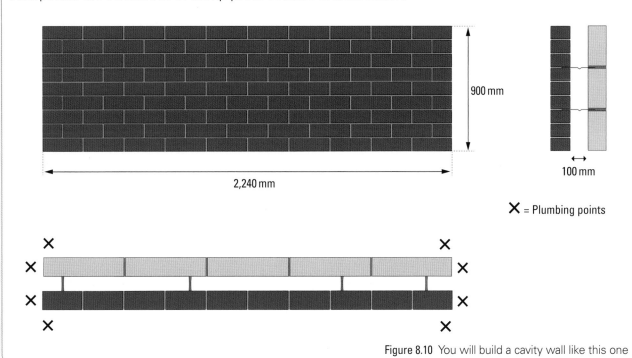

900 mm

2,240 mm

100 mm

✕ = Plumbing points

Figure 8.10 You will build a cavity wall like this one

223

TOOLS AND EQUIPMENT

Walling trowel

Lump hammer

Bolster chisel

Spirit level

Builder's square

Block/pins and line

Straight edge

Jointing iron

Steel tape measure

PPE

Ensure you select PPE appropriate to the job and site where you are working. Refer to the PPE section of Chapter 1.

STEP 1 Dry bond 10 bricks to a length of 2,240 mm. Lay the first brick at the highest end of the floor and then transfer the level from one end to the other with a spirit level and straight edge.

PRACTICAL TIP

Remember to rotate the level 180° in order to eliminate any possible inaccuracies in the level if using this method.

STEP 2 Check the alignment with the straight edge and then re-check that length, level and alignment are all correct before starting walling.

STEP 3 Place two or three bricks onto the top of the two end bricks to stop them from moving when the lines are fitted with the blocks.

STEP 4 The bricks can then be laid in to complete the first course. This should provide an accurate starting point from which all other parts of the model can be established.

STEP 5 Repeat the setting out procedure with the blocks so that the block part of the wall is also 2,240 mm long and directly opposite the outer leaf, with the correctly sized cavity (100 mm).

PRACTICAL TIP

Check the blocks for upright with a level on the first course as this will make plumbing of further courses easier.

STEP 6 Build up the block corners with the ties placed in the correct position.

PRACTICAL TIP

Ensure that the ties are added at every course on the ends of the cavity between 150 mm and 250 mm from the end, depending on the type of cavity closer.

PRACTICAL TIP

Lay the ties out on the wall before laying the bed to reduce the risk of forgetting to bed them in.

STEP 7 Run in the block wall to full height with the line. Place wall ties vertically at 225 mm from either of the open ends of the cavity, and at 450 mm heights throughout the body of the wall. Arrange them in a diamond pattern as much as possible, so that the ties are taking equal loads.

STEP 8 Repeat the procedure on the brick leaf by building small corners and run in the courses to the line.

STEP 9 Apply a half-round joint to the brickwork and finish the blockwork with a flush joint.

PRACTICAL TASK

2. BUILD A STRAIGHT CAVITY WALL WITH INSULATION

OBJECTIVE

To practise building a cavity wall and adding insulation to the required specifications.

This is the same as the previous task, only this time insulation has been added. This is of the partial fill type of construction. Therefore, a clear 50 mm gap has to be maintained between the insulation and the outer leaf of the brickwork. For this reason, the cavity should be constructed at 100 mm as this is determined by the 50 mm thickness of the insulation.

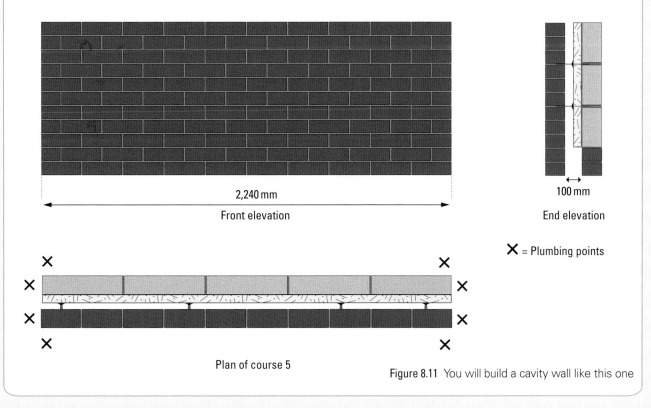

2,240 mm
Front elevation

100 mm
End elevation

X = Plumbing points

Plan of course 5

Figure 8.11 You will build a cavity wall like this one

TOOLS AND EQUIPMENT

Walling trowel

Lump hammer

Bolster chisel

Spirit level

Builder's square

Block/pins and line

Straight edge

Jointing iron

Steel tape measure

Knife or scissors

PPE

Ensure you select PPE appropriate to the job and site where you are working. Refer to the PPE section of Chapter 1.

STEP 1 Follow Steps 1 to 4 of the previous practical exercise.

STEP 2 Build the second course in order to get to DPC height (150 mm). You can then use this to determine the position and height of the two courses that will form the inner leaf. These must also be two courses high so that the subsequent brick and block courses are at the correct heights in relation to each other.

Figure 8.12 Getting to DPC height

STEP 3 A thin screed of mortar should be laid out and the DPC applied. This mortar is not to stick the DPC, but rather to protect it from any high spots or sharp areas that may puncture it and compromise its efficiency. Any overlaps should be a minimum of 100 mm.

Figure 8.13 Adding a thin screed of mortar

Figure 8.14 Attaching the DPC

Figure 8.16 Close up of ties

STEP 4 Build the block corners first.

STEP 5 Horizontal spacing should be 900 mm maximum and the ties at the stop ends should be placed a minimum of 150 mm from the end, so as not to interfere with any methods of closing the cavity.

PRACTICAL TIP

It may help to position the wall ties on top of the DPC to support the insulation. On site, it may be that the insulation protrudes below DPC and the ties will be applied accordingly.

STEP 6 Now apply the insulation. You may have to reduce the horizontal insert of the wall to support the insulation. Cut it to fit so that it is 'bonded' in much the same way as stretcher bond is. Cut it at the ends so that any cavity closers can be fitted without having to alter the insulation. The amount this should be cut depends on the depth that the closer sits in the cavity, but 50 mm should be sufficient. Then fix the insulation with the appropriate plastic clips.

STEP 7 Now the brickwork can be constructed. As usual, start by building small corners and then run them in with the line in order to gain the best efficiency.

STEP 8 Finish the wall with a half-round joint on the brickwork and a flush finish on the block leaf. Brush the wall when the mortar is dry enough for this not to leave any brush marks.

Figure 8.15 The fixed insulation

3. BUILD A CAVITY WALL WITH A JUNCTION

OBJECTIVE

To build a cavity wall with a return added to the inner skin of blockwork.

On an actual construction this can be seen where party walls are built of the main construction and form the internal walls. In this model the return is bonded into the inner block leaf. If separate materials were used, however, modern practice is to fix the partition and walls with fixings as this prevents cracking due to differing thermal expansion rates between the materials.

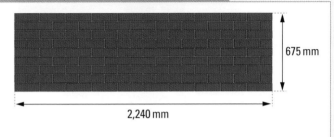

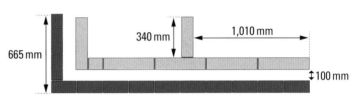

Figure 8.17 You will build a cavity wall like this one

TOOLS AND EQUIPMENT

Walling trowel	Jointing iron
Lump hammer	Steel tape measure
Bolster chisel	
Spirit level	
Builder's square	
Block/pins and line	
Straight edge	

PPE

Ensure you select PPE appropriate to the job and site where you are working. Refer to the PPE section of Chapter 1.

STEP 1 Dry bond 10 bricks 2,240 mm in length. Lay the first brick at the highest end of the floor and then transfer the level from one end to the other with either a spirit level or straight edge.

STEP 2 Check the alignment with the straight edge and then re-check all that length, level and alignment are all correct before starting walling.

STEP 3 Place two or three bricks onto the top of the two end bricks to stop them from moving when the lines are fitted with the corner blocks.

STEP 4 Lay the bricks in to complete the first course. This should provide an accurate starting point from which all other parts of the model can be established.

STEP 5 The return corner can then be established to the correct length and checked for 90° with either a builder's square or the 3:4:5 method (see page 203). This can then be repeated with the block leaf, making sure that the cavity is a constant 100mm wide.

STEP 6 Position the return block and check for level and plumb. Check that the block is 90° to the block leaf.

PRACTICAL TIP

As the line will not be there to check the bottom of the blocks on the first course, use a level to plumb them in to make the laying of subsequent courses easier.

STEP 7 The block corners should be built first, making sure that the ties are correctly fitted.

STEP 8 When running in the blockwork, leave the return until last on each course and then check it for alignment with a spirit level.

Wall ties should be placed 225mm vertically at either of the open ends of the cavity, and at 450mm heights throughout the body of the wall, and placed in a diamond pattern as much as possible, so that the ties are taking equal loads. Horizontal spacing should be a maximum of 900mm and the ties at the stop ends should be placed a minimum of 150mm from the end, so as not to interfere with any methods of closing the cavity.

STEP 9 Now construct the brickwork. As usual, start by building small corners and then run in with the line in order to gain the best efficiency.

STEP 10 The wall should be finished with a half-round joint on the brickwork and a flush finish on the block leaf. Brush the wall when the mortar is sufficiently dry for this to not leave any brush marks.

Figure 8.18 Building the partition wall

Figure 8.19 The wall with the partition built

Figure 8.20 The wall showing the cavity and return

TEST YOURSELF

1. What is another term used to describe the walls that make up cavity walling?

   a. Leaves

   b. Brickwork

   c. Blockwork

   d. Cavity spacer

2. What is the bend or kink in a wall tie known as?

   a. Droplet

   b. Dropper

   c. Drip

   d. Weep hole

3. How many brick lengths is a cavity tray over an air brick?

   a. 1

   b. 2

   c. 3

   d. It varies

4. What is the joint finish called where the joint slopes downwards to encourage rainwater away from the joint?

   a. Weather struck

   b. Flush

   c. Recessed

   d. Ironed

5. Bridges can be formed between cavity walls. Which of the following is **not** a way that a bridge can be formed?

   a. Mortar droppings inside the cavity

   b. Poor placement of wall ties

   c. Problems with the damp-proof membrane

   d. The air in the cavity heating up

6. Ideally, how often should wall ties be spaced in brickwork at an opening?

   a. Every second course

   b. Every third course

   c. Every fourth course

   d. Every fifth course

7. How deep should wall ties be bedded into each wall leaf?

   a. 25mm

   b. 50mm

   c. 75mm

   d. 100mm

8. Which of the following statement is **incorrect**?

   a. Wall ties should be pressed down firmly into the mortar bed

   b. You should never push ties into joints

   c. The drip needs to be pointing downwards and be at the centre of the cavity

   d. Wall ties can be bent to match mortar beds

9. Which of the following statements is **true**?

   a. On hot days joints will dry out slowly

   b. The softer the brick the more moisture it will soak up

   c. Brick courses laid earlier in the day will be wetter

   d. On cold and wet days joints quickly dry out

10. What type of bond is the most commonly used for both inner and outer leaves of cavity walls?

    a. Stretcher bond

    b. Broken bond

    c. Reverse bond

    d. Flemish bond

INDEX

ACKNOWLEDGEMENTS

The author and the publisher would also like to thank the following for permission to reproduce material:

Images and diagrams

Alamy: Arcaid Images: 3.12, blickwinkel: 3.13, Peter Davey: chapter 1 opener; **BSA**: 2.5; **Energy Saving Trust © 2013**: 3.28; **Fotolia**: 1.1, 1.2, 1.3, 1.5, 1.6, 1.7, 1.8, 1.14, 1.15, 1.16, 7.2; **Getty Images**: Fotosearch Value, chapter 6 opener; **Helfen**: 2.4; **instant art**: table 1.15; **iStockphoto**: 1.11, 2.26, 3.4, 3.16, 3.18, 3.19, 3.21, 3.22, 3.25, 3.26, 3.27, chapter 5 opener, 6.5, 6.20; **Nelson Thornes**: 1.9, 1.10, 1.12, 1.13, 4.10, 4.11, 4.12, 4.13, 4.14, 4.15, 4.16, 4.17, 4.18, 4.19, 6.1, 6.2, 6.3, 6.4, 6.6, 6.9, 6.10, 6.11, 6.12, 6.13, 6.14, 6.15, 6.16, 6.17, 6.18, 6.19, 6.21, 6.22, 6.23, 6.24, 6.25, 6.26, 6.27, 6.28, 6.29, 6.30, 6.31, 6.33, 6.34, 6.35, 6.36, 6.37, 6.38, 6.39, 6.4, 6.41, 6.42, 6.44, 6.45, 6.46, 6.47, 6.48, 6.49, 6.5, 6.53, 6.54, 6.55, 6.56, 6.57, 6.58, 6.59, 6.61, chapter 7 opener, 7.3, 7.11, 7.13, 7.29, 7.30, 7.31, 7.32, 7.33, 7.34, 7.35, 7.37, 7.38, 7.39, 7.41, 7.42, 7.43, 7.44, 7.45, 7.46, 7.47, 7.48, 7.49, 7.5, 7.51, 7.52, 7.53, 7.54, 7.56, 7.57, 7.58, 7.59, 7.6, 7.61, 7.62, 7.63, 7.64, 7.65, 7.66, 7.67, 7.68, 7.69, 7.71, 7.72, 7.73, 7.74, 7.75, 7.76, 7.77, 7.78, 7.8, 7.81, 7.82, 7.84, 7.86, 7.87, 7.89, 7.91, 7.93, 7.94, 7.95, 7.96, 7.97, 7.98, 7.99, chapter 8 opener, 8.12, 8.13, 8.14, 8.15, 8.16, 8.18, 8.19, 8.20; **Peter Brett**: 2.1, 2.3, 2.6, 2.8; **Science Photo Library**: Peter Gardiner: 1.4; **Shutterstock**: chapter 2 opener, chapter 3 opener, 3.11, 3.14, 3.20, 3.23, 3.24, chapter 4 opener, 5.1, 7.1; **Solent Building Supplies Ltd**: 7.10; **STA**: 7.24; **Wikipedia**: 3.15.

Every effort has been made to trace the copyright holders but if any have been inadvertently overlooked the publisher will be pleased to make the necessary arrangements at the first opportunity.